AF564561

Recent Developments in Plant Breeding

NIPA® GENX ELECTRONIC RESOURCES & SOLUTIONS P. LTD.
New Delhi-110 034

About the Editors

Dharm Veer Singh is perusing his Ph.D degree at Department of Genetics and Plant Breeding, College of Agriculture, Banda University of Agriculture and Technology, Banda, Uttar Pradesh, India. His research work is focused on the identification of heat-tolerant lentil genotypes through biochemical and transcriptomic approach.

Dr. Kamaluddin is a Professor in the Department of Genetics and Plant Breeding, College of Agriculture, Banda University of Agriculture and Technology, Banda, Uttar Pradesh, India. He completed Ph.D from Institute of Agriculture Science, Banaras Hindu University, Varanasi. He has expertise in plant breeding, transgenic technology, and molecular breeding. Dr. Kamaluddin has published original research articles, review articles, book chapters, and abstracts in different journals of national and international reputes. Presently, he is involved in the development of heat and drought stress-tolerant pulse cultivars using conventional and marker-assisted selection. Further, he is also working on development of varieties especially in small millets. He has published *Plant Biotechnology: Principles and Application* (Springer), *Transgenic Technology Based Value Addition in Plant Biotechnology* (Elsevier) as coeditor, and *Technologies in Plant Biotechnology and Breeding of Field Crops* (Springer) as Editor.

Dr. Vijay Sharma is currently working as Assistant Professor at Department of Genetics and Plant Breeding, College of Agriculture, Banda University of Agriculture and Technology, Banda, Uttar Pradesh, India. His research work is focused on development of varieties in sesame and wheat.

Dr. Shiva Nath is working as Associate Professor at Acharya Narendra Dev University of Agriculture and Technology, Kumarganj (Ayodhya), Uttar Pradesh, India. He is currently working on chickpea breeding.

Recent Developments in Plant Breeding
Volume 1

Dharm Veer Singh
Department of Genetics and Plant Breeding
College of Agriculture
Banda University of Agriculture and Technology
Banda, Uttar Pradesh, India

Kamaluddin
Department of Genetics and Plant Breeding
College of Agriculture
Banda University of Agriculture and Technology
Banda, Uttar Pradesh, India

Vijay Sharma
Department of Genetics and Plant Breeding
College of Agriculture
Banda University of Agriculture and Technology
Banda, Uttar Pradesh, India

Shiva Nath
Acharya Narendra Dev University of Agriculture and Technology
Kumarganj, Ayodhya, Uttar Pradesh, India

NIPA® GENX ELECTRONIC RESOURCES & SOLUTIONS P. LTD.
New Delhi-110 034

NIPA® GENX ELECTRONIC RESOURCES & SOLUTIONS P. LTD.

101,103, Vikas Surya Plaza, CU Block
L.S.C. Market, Pitam Pura, New Delhi-110 034
Ph : +91-11-43860225, Mob.: +91 9717133558, 9540816132
E-mail: newindiapublishingagency@gmail.com
Website: www.nipaersources.com

Print ISBN: 978-93-58872-75-0
ebook ISBN: 978-93-58878-62-2

NIPA® also publishes books in a variety of electronic formats. Some content that appears in print may not be available in electronic books, and vice versa.

Composed and Designed by NIPA®.

Preface

As the effects of climate change become increasingly apparent and inescapable, human society is at a tipping point at present. The effects of climate change are unparalleled worldwide, ranging from rising temperatures that threaten food production to melting glaciers that result in catastrophic flooding and erosion. Geneticists and plant breeders remain under pressure to sustain food production through revolutionary breeding techniques and the improvement of minor crops that are resistant to both biotic and abiotic challenges and are well-suited to the marginal soil and source of nutrition. The modern and innovative technologies used by breeder may save mankind from impending agricultural problems brought on by changing weather patterns, pests that are emerging quickly, and finite resources. Every crop development initiative must, invariably, unlock the reservoir of genetic diversity and make full use of wild germplasm. But present advances in high-throughput phenomics, breeding strategies, genomics, and genome editing, bring up new avenues for accelerating the improvement of crops suitable in climate change scenario.

In order to achieve sustainable agricultural production and improved food security, this book discusses the development of innovative modern methodologies to supplement conventional plant breeding for the production of new crop varieties under the increasingly restrictive environmental and cultivation component. The chapters provide in-depth information as well as current research on certain selected topics that will help understand the recent developments in plant breeding. Plant breeders and geneticists engaged in breeding assignments integrating traditional breeding with biotechnology and molecular breeding methods and also advanced undergraduate and graduate students will find the book to be a valuable resource.

The book is divided into seventeen chapters, each of which summarizes and replicates the author's own understanding of the pertinent information.

Chapter 1: Introduction covers important topics of crop domestication, pre-Mendelian, Mendelian, and post-Mendelian plant breeding, and relation of plant breeding with other disciplines of agriculture.

Chapter 2: Modern and Transgenic Breeding Approaches describes the modern and molecular approaches of cultivar development. Various advanced molecular tools, including modern plant-breeding triangle, high-throughput phenotyping, marker-assisted selection, marker-assisted backcrossing, genomic selection, genome-wide association study , and genome editing, have been deliberated in this chapter, which facilitates the knowledge flow from scientists to farmers.

Chapter 3: Breeding Strategies for Disease Resistance highlights the various breeding strategies employed for speedily improving plant tolerance to diseases. This chapter also describes marker-assisted selection, genomic selection, and gene editing expedite targeted trait acquisition and gene pyramiding. By combining cutting-edge technology with conventional knowledge, these techniques strengthen plants' ability to fight back against diseases, ensuring environmentally friendly agriculture and an adequate supply of food for mankind.

Chapter 4: Breeding Strategies for Insect Resistance ....

Chapter 5: Breeding Strategy for Salinity Stress Tolerance describes the major conventional and modern tools used for developing varieties tolerance against salt stress.

Chapter 6: Plant Tissue Culture: Application in Crop Improvement describes historical background of plant tissue culture, its present and future scenario, different methods and application of plant tissue culture in plant breeding.

Chapter 7: Breeding Approaches in Transgenic Crops: Ethical Issues and Regulation attempts to explain different approaches in transgenic development, its benefit in crop improvement and ethical issues and regulatory framework related to transgenic crops.

Chapter 8: Marker-assisted Selection: Approaches and Importance in Plant Breeding describes different approaches of marker-assisted selection, steps involved in it, its different approaches using in crop improvement, and importance in plant breeding.

Chapter 9: Genome Editing: Approaches, Principles and Application in Plant Breeding explores genome editing technologies, with a focus on CRISPR-Cas9, TALENs, and ZFNs. The chapter has also highlighted applications of genome editing in plant breeding.

Chapter 10: Advance Genomic Technologies: TILLING and Eco-TILLING describes the steps involved in TILLING and Eco-TILLING along with applications and merits and demerits.

Chapter 11: Advancements in Forward and Reverse Genetics highlights about the forward and reverse genetics. This chapter also covers the previous techniques of forward genetics and reverse genetics and their limitations, recent techniques of forward and reverse genetics and their advantages and disadvantages over traditional techniques.

Chapter 12: Pre-breeding: Its Application in Plant Breeding describes pre-breeding, why it is required along with objectives, strategies, and application and challenges involved.

Chapter 13: Role of Wide (Distant) Hybridization in Crop Improvement: Applications and Limitations covers wide hybridization and its different types, pre- and post-fertilization barriers and techniques to overcome these barriers, its role, limitations, achievements, and future prospects in crop improvement.

Chapter 14: Speed Breeding: Importance and Future Prospects in Agriculture describes speed breeding and its strategies involved in crop improvement. This chapter has also highlighted the advantages, challenges, future outlook, and significance of speed breeding in cultivar development.

Chapter 15: The Role of Epigenetics in Plant Breeding highlights history and evolution of epigenetics, methods, and mechanism of epigenetic modification, different epigenetic mechanism, need of epigenetics in plant breeding, and exploitation of epigenetic diversity and its significance.

Chapter 16: Concepts and Models of Stability Analysis in Plant Breeding describes genotype–environment interaction and its significance, different stability analysis models, and their advantages and disadvantages.

Chapter 17: Intellectual Property Right and its Requirement in Plant Breeding attempts to elucidate historical concept of IPR, IPR system in the world and India, and forms and scope related to agriculture.

It has been incredibly tough to compile and update this book while working full-time jobs as researchers. We greatly appreciate the support of our friends, family, and students. The book comes as a great resource to graduate and postgraduate students, teachers, plant breeders, researchers, policymakers, and readers in general, who have an interest in agriculture, especially with regard to modern breeding technologies.

Editors

Contents

1

Introduction

***Kamaluddin*[1]*, Vijay Sharma*[2]*, Shiva Nath*[3] *and Dharm Veer Singh*[4]**

[1]*Professor, Department of Genetics and Plant Breeding, Banda University of Agriculture and Technology, Banda, Uttar Pradesh, India*

[2]*Associate Professor, Department of Genetics and Plant Breeding, ANDUA&T Kumarganj, Ayodhya, India*

[3]*Assistant Professor, Department of Genetics and Plant Breeding, Banda University of Agriculture and Technology, Banda, Uttar Pradesh, India*

[4]*Ph.D. Research Scholar, Department of Genetics and Plant Breeding, Banda University of Agriculture and Technology, Banda, Uttar Pradesh, India*

Abstract

This chapter covers plant breeding's present, past, and future prospects with a special emphasis on cultivar development. In order to increase agricultural production and productivity (yield per unit area) of important crops, plant breeders have achieved significant achievements in the genetic enhancement of yield, adaptive features, disease–pest resistance, abiotic stress tolerance, and nutritional qualities. Plant breeding is more crucial than ever to address production issues in order to provide a steady supply of food, feed, and fiber due to the growing demands of the human population and rising climate fluctuations. Now a days plant breeders have advanced tools to create superior cultivars because of recent technological developments in the fields of phenomics, genomics, and biotechnology. This chapter has covered important topics of crop domestication, pre-Mendelian and post-Mendelian plant breeding, plant breeding in the 20th and 21st centuries, including information on the green revolution, genetically engineered crops, molecular marker applications, image-based phenotyping, machine learning techniques, and the relationship between both public and private plant breeding organizations, and future plant breeding efforts.

Keywords: *Genetic, cultivar, domestication, phenomics, biotechnology*

Introduction

With respect to altering the genetic composition of plants for the global economic benefit, plant breeding is defined as "an art and science" and a technology.

Plant breeding is a conscious attempt to positively influence nature with regard to the plants' genetic constitution. The modifications introduced to plants are both permanent and inherited.

Ten thousand years ago, hunter-gatherers started to specialize in gathering and tending to a small number of favorite plant species in western Asia, and later in eastern Asia, the Americas, and Africa. The preferred species were eventually deliberately planted in separate areas, the most desired plants' seeds were transplanted and gradually managed crops were developed.

Plant Breeding is an Art and Science

The expertise of the breeder in identifying plants with distinctive monetary, environmental, nutritive, or aesthetic properties is the essence of plant breeding. Before having access to the scientific information that they do now, plant breeders were limited to using their expertise and intuition to choose novel plants that might be reproduced by seed or vegetative parts. As a result, selection originated as the first method of plant breeding. In order to increase efficiency in the field or garden, effective plant breeders have to be excellent observers and fast to identify variations of the same kind of plants. Plant breeding was only an artistic endeavor to them. Many of the earlier breeders were inexperienced cultivators who discovered an "off-type" plant in the field or a "sport" in a bed by accident.

As understanding of classical genetics and allied plant sciences increased, plant breeding became a science. The principles of genetic behavior that made it possible to predict precisely the outcomes of gene manipulations, methodologies for manipulating genes, and mention of the gene as the basic unit of heredity served as the foundation for plant breeding. The observable appearance of features in plants, such as height or (Tall and dwarf) or the color of the bloom (white or pink), was used to identify the genes. Specific gene combinations for several outstanding features might be merged into a single plant cultivar by disciplined cross-pollination. The main method for plant breeding was hybridization. The breeder no longer had to depend so much on expertise in order to identify chance variations with which to create new cultivars. New plant varieties might now be planned and created almost at will. Plant breeding shifted from being an art to being more of a science.

The study of molecular genetics has currently made approaches to take plant breeding to a new degree of complexity. The description of the chemical composition of deoxyribonucleic acid (DNA), the substance that makes up a gene, marked the beginning of molecular genetics. Specific enzymes and proteins that control how some plant features manifest themselves visibly are encoded in DNA. A desired trait's DNA (gene) would be detected, replicated, and incorporated into a plant reproductive cell line in accordance with the new technique. As the new technique becomes more widely used, it presents a chance to improve agricultural cultivar performance by introducing foreign genes from almost infinite sources, genes that were not previously available through conventional hybridization breeding processes.

Promising implications for plant breeding result from the improvement of cultivar performance by transformation using single DNA units (genes), which encode proteins that describe alien plant characteristics. However, the old selection and hybridization processes used in the breeding of better cultivars are still necessary and do not get replaced by the new technology, nor does it lessen their significance or requirement. Through sophisticated selection and hybridization breeding techniques that combine many genetic combinations for plant performance, the current cultivars have attained high levels of performance. High-performing cultivars will continue to be mainly produced by these breeding techniques, however, performance may be enhanced further by novel biotechnology approaches.

Why Breeding is Necessary?

Plant breeding aims to alter a plant's genetic makeup in a way that will enhance its performance.

There are several ways that improved plant performance might be seen. The main breeding objectives are often increased production and quality, regardless of whether the product is collected as seed, fodder, fiber, fruit, tubers, flowers, or other plant components. The primary source of sustenance for everyone on earth is plants. Higher food plant yields result in a more plentiful food supply, more lucrative agriculture, and more affordable food prices for consumers. Over the 35-year period from 1965 to 2000, the principal food grain yields in the United States expanded quickly. The improved cultural conditions in which the crops were produced and the genetic capacity of new cultivars to produce under these conditions led to an increase in yield. Food plants that have been bred for higher quality may be more nutrient-dense, easier to digest, or contain less harmful substances. As less protective chemicals will be needed in cultivation of the resistant plants, breeding plants for disease or insect resistance improves plant health while also increasing yield and product quality. By breeding for greater tolerance to elevated temperatures, drought, salt, or other harmful environmental production risks, plants can be adapted to a wider range of production environments.

Plant Breeding Strategic Planning

Plant breeding techniques are comparatively simple. The fundamental components of this strategy are to:

- Identify the morphological, physiological, and pathological characteristics in cultivated plant species that influence their adaptability, health, productivity, and suitability for food, fiber, or industrial products.
- Seek for new genes that encode desired traits in various strains of the cultivated species and their close relatives.
- Combine genes for the desired traits into an improved cultivar through traditional breeding or current biotechnological approaches.
- Evaluate the performance of the enhanced breeding lines in the local environment in comparison with the existing cultivars and disseminate as new cultivars breeding lines superior to cultivars currently farmed.
- How this method is now implemented will be addressed in subsequent chapters.

Advancement of Modern Plant Breeder

The learners could inquire, "What do I study in order to become a plant breeder?" The easiest response is "You must examine plants," yet this requires studying a variety of academic fields. The education of the modern plant breeder must include knowledge in several areas of plant science (Fig. 1).

They consist of the following:

- ***Botany***: In-depth knowledge of the taxonomic categorization, anatomy, morphology, reproductive system, and cellular structure of the agricultural plants they work with is essential for plant breeders. They should also be skilled biologists.
- ***Genetics***: Modern plant breeding techniques are founded on knowledge of the gene and its inheritance, thus a plant breeder must have a full understanding of the mechanisms of heredity in plants. The understanding of genes has been expanded to the molecular level because of developments in molecular genetics.

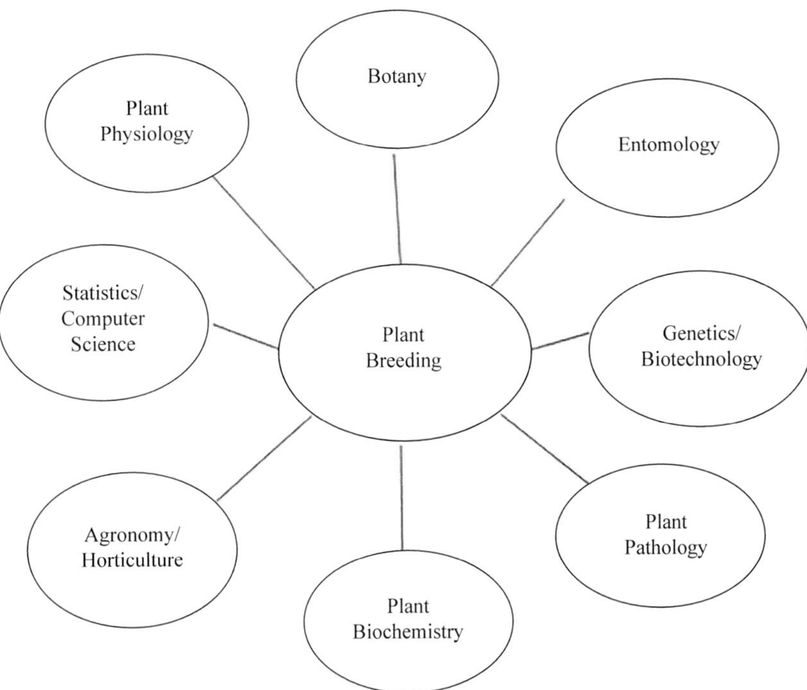

Fig. 1: Plant breeding and other plant science areas are interconnected to each other. The breeder has to have a solid understanding of various fields of plant science in order to create a better crop variety. The extraordinary successes in plant breeding are typically the outcome of the breeder's collaboration with experts in these fields.

- ***Plant physiology*:** Plant responses to environmental challenges, such as peaks in sunlight, temperature, moisture levels in the soil, and soil nutrients, have an impact on cultivar adaptability. The goal of a plant breeder is to alter a plant's physiological systems so that it can operate more effectively in its current environment.
- ***Plant pathology*:** Strong plants are necessary for successful crop production. The discovery of genes for resistance to plant pathogens that cause diseases is a collaborative effort between the plant pathology and the plant breeder. The performance of plants is enhanced and the need for chemical disease control is decreased when disease-resistant genes are incorporated in cultivars.
- ***Entomology*:** Reducing the use of pesticides in agricultural and horticultural practices and breeding for insect resistance is a cost-effective and ecologically responsible way to prevent insect damage.

- ***Plant biochemistry***: Plant breeding frequently has the potential to enhance a crop cultivar's natural nutritional value for use as food, livestock feed, or industrial input. Examples include tomato texture and flavor, higher lysine content in food grains, wheat's milling and baking quality, or cotton's fineness and strength of fiber. The chemical makeup and purpose of genetic material are now better understood because of molecular genetics.
- ***Statistics***: In the breeding nursery, the performances of strains with comparable genetic makeup are compared. Analytical statistical methods help to increase knowledge about quantitative genetics and its application in plant breeding.
- ***Computer science***: For precise design of the breeding nursery, collection of observations, and quick analysis and interpretation of the data, the computer has become a necessary instrument.
- ***Horticulture and agronomy***: Breeders must be knowledgeable about crops and how to grow them. In order to analyze the available breeding materials, establish effective breeding processes, and focus breeding efforts on significant breeding goals, they should be aware of the grower's requirements for new cultivars of field or horticultural crops.

All of these areas of plant science are outside the scope of a plant breeder. The breeder is not engaged in any one of them entirely when engaging in the practice of plant breeding. The job of the plant breeder is to use all of his plant-related knowledge and expertise to create better cultivars.

The breeder may start research to investigate those specific issues if more knowledge is required regarding the inheritance of a plant feature or regarding a method for comparing the tolerance of several cultivars in a given environment.

It is ideal to conduct joint research projects with researchers in relevant areas in order to tackle shared difficulties in addition to a breeding program. The fastest progress is accomplished when a team of geneticists, pathologists, physiologists, entomologists, or biochemists collaborates with the plant breeder because the genetic development of plant species requires study in many departments of science. The outcomes of such collaboration in plant breeding are frequently remarkable successes.

How Plant Breeding Started?

Plant breeding began when prehistoric humans discovered how to find superior plants to harvest. The species of agricultural plants that are grown today, with very few exceptions, developed from wild predecessors over a long period of time. The early practice of gathering mutant plants with beneficial features accelerated the domestication of wild species. An excellent example is wheat, which was domesticated after deriving from wild progenitors that were quite different from contemporary varieties.

The seeds of the early wheat fell to the soil surface as they matured because the seed-bearing spikes were fragile and readily broken. Before seeds could be consumed, they were protected by hulls that stuck to them securely and were challenging to remove.

Over extended periods of time, may be 5,000–10,000 years, mutant varieties of wheat slowly developed that allowed seeds to remain securely in place until harvest while yet being easily separated from hulls. Because they were simpler to grow and prepare for food, these new varieties of wheat were harvested. They eventually evolved into the wheat that people in the distant past farmed.

Indian maize (maize), which the early Native Americans domesticated, was a great achievement.

Indian corn's origins are less specific than the ancestors of wheat, but it appears to be quite certain that the ancient maize possessed small, flinty kernels similar to our modern popcorn. Native American tribes who lived in various climate zones made radically diverse plant and kernel selections as maize domestication and development continued. The "races" of maize that have evolved have a variety of endosperm types, from small-seeded, flinty varieties found in North America's central plains to large-seeded, floury varieties grown in Central and South America.

The Native Americans who lived on the high plateaus of Bolivia and Peru are thought to be responsible for domesticating the potato. Wild potato tubers frequently included alkaloids, which gave them a bitter flavor and rendered them fairly hazardous if consumed. Mutant forms that were less poisonous and bitter were collected as they were found, and they eventually evolved into the ancestors of today's farmed potatoes.

Other domesticated plants, including cotton, tobacco, tomato, and capsicum pepper in the Americas; barley, oats, and chickpea in the Middle East; rice and sugarcane in southeast Asia and nearby South Pacific islands; soybean in China; sorghum and millet in Africa, and many others, have similar origins and domestication stories from wild species by early humans.

Even today, wild species are still being cultivated for food, generally by a competent plant breeder who genetically modifies the wild plants to increase their usefulness and productivity. Through deliberate plant breeding programs, weedy wild plants like the sunflower and beet root were transformed into cultivated crops. Early varieties of beetroot united flower clusters from a single axil without cleavage walls, resulting in multigerm seedball development.

A monogerm seedball, which produces a single seedling, is created by breeding new cultivars with a sole flower in the axil of a bract. This technique eliminates the need for hand thinning to provide uniform distance between sugar beet plants.

Domestication of Wild Rice (Zizania Palustris)

This started only lately as a direct outcome of intentional plant breeding initiatives in the northern United States. Native Americans had been gathering wild plants growing in small lakes in northern Minnesota and neighboring parts of Canada for sustenance over many years. The fully mature seeds break when they are grown, fall to the lake's bottom where they overwinter, and then sprout to begin the harvest for the next season. Native Americans would often float among the upright plants and gather the seeds by shaking them into their canoes.

Wild rice has become more popular as a gourmet meal in recent years to the point that the meager harvest using traditional techniques from native stands failed to keep up with the demand. Nonshattering cultivars were created through a plant breeding program.

Plan Breeding Prior to Mendel

It would be challenging to pinpoint the exact moment when humans first deliberately started breeding plants, but certain significant early milestones in the development of selection and hybridization may be noted. The progeny test was used in 1856 by a Frenchman named Louis Leveque de Vilmorin to raise the sugar content of the wild beetroot. The progeny test measures a plant's breeding value based on the performance of its offspring. The success of creating cultivars from the offspring of a single plant in self-pollinated species was established at the Swedish Seed Association's plant breeding facility at Svalöf just before 1900 (Roll-Hansen 1986). Professor Hjalmar Nilsson proved that the plant is the proper selection unit, and not just an individual flowering structure as some breeders had previously suggested during that time.

In 1913, Wilhelm Johannsen, a Danish botanist, used studies on garden beans to demonstrate the correctness of utilizing the plant as the unit of selection to create uniform, true-breeding cultivars, while Willet M. Hays, an American botanist, used wheat to demonstrate the correctness of the same method.

W.A. Orton reported groundbreaking achievements in the field of breeding for disease resistance in cotton wilt resistance in 1898, while H.L. Bolley reported similar achievements in linseed wilt resistance in 1901. By growing the plants in soil that was infected with the wilt bacterium and then selecting the surviving plants, the two researchers exposed mixed populations of plants to a natural epidemic of the illness.

Today, breeding for resistance to pathogens that cause plant disease is fundamentally based on the survival principle. Awareness of basic hereditary principles and their application in the logical development of hybridization as a plant breeding technique need knowledge of the role of pollen in the fertilization of plants as well as an awareness of the plant's reproductive system. R.J. Camerarius, a German botanist, conducted studies that made it possible to conclusively prove the existence of sex in plants as early as 1694, but it took until the late 1800s for gamete fusion and double fertilization to be fully understood.

Roles of Gregor Johann Mendel

Prior to 1900, there were many plant breeders working, but none had an impact comparable to that of Augustinian monk, Gregor Mendel. Mendel cross-pollinated multiple varieties of the common garden pea while working in a monastery garden in Brünn, Bohemia, and he observed how the ratios of the various varieties changed over time. Mendel developed a few key "Laws of Inheritance" via careful observation and deft reasoning. These laws remain true today and serve as the foundation for plant breeders' procedures. Mendel's discovery was originally published in 1865, but it was not recognized until three plant researchers independently confirmed it in 1900. Since then, an extensive amount of new information has been added to and expanded upon Mendel's original Laws of Inheritance. This knowledge is a component of the vast and significant field of study known as genetics. In addition to any other branch of study, genetics is the basis on which plant breeding has evolved.

Plant Breeding after Mendel

A large number of plant scientists engaged in plant breeding research were inspired by Mendel's studies to further their work. The technique for breeding hybrid maize was found because to one of the investigations.

At the Station for Experimental Evolution in Cold Spring Harbor, New York, G.H. Shull began breeding open-pollinated maize in order to conduct the study on heterosis breeding in maize in 1904.

Because hybrid maize was so productive, "open-pollinated" maize that was being cultivated at the time was eventually replaced. Since then, a variety of field and horticultural crops, including tomato, pearl millet, sorghum, sunflower, tobacco, and sunflower, have been bred using hybrid vigor.

High levels of cultivar performance are usually achieved by the practice of fusing genes from different sources of germplasm. Short, profusely tillering, and extremely productive wheat cultivars that increased wheat yields to a new high were later developed by crosses between short "semidwarf" wheat from Italy and Japan and locally grown wheat. By using the new germplasm to create a number of short-statured, highly productive cultivars, that became the foundation for a "green revolution" in wheat production when transferred into subtropical regions of the world, Norman E. Borlaug internationalized wheat breeding. The high-yielding wheat Borlaug developed contributed to the increase in global food supply, and he was rewarded the Nobel Peace Prize in 1970 for his efforts.

The International Rice Research Institute, Los Baos, Laguna, Philippines, created short-statured rice cultivars under the direction of Dr. T.T. Chang, which served as the foundation for the green revolution in rice production. These cultivars improved rice yields by >40% in tropical Asia.

T.J. Jenkin, whose idea of strain building was established at the Welsh Plant Breeding Station, Aberystwyth, starting in 1919, produced significant improvements in forage crop breeding techniques.

Cross-pollinated species' individual plants are chosen and incorporated in a breeding process known as strain development to create a synthetic cultivar. Today, numerous forage species, including lucerne, are improved using this method.

Glen W. Burton showed that it was feasible to choose a fruitful hybrid plant of Bermuda grass, a species that produces seed seldom, and propagate the hybrid cultivar by sowing vegetative sprigs at the Coastal Plains Experiment Station in Tifton, Georgia. Numerous further advancements in fodder breeding were made possible by Burton's original concepts. In the tropical plant pearl millet, he discovered cytoplasmic male sterility and fertility-restorer genes and developed a method for producing hybrids of that species.

Edgar E. Hartwig spent 50 years in Stoneville, the state of Mississippi where he produced numerous cultivars of soybean renowned for their production and their resistance to pests, worms, and diseases.

Conclusion

Plant breeding has a lot of capacity to support secure food supply and sustainable agricultural development (Qaim 2020). Over the past 100 years, plant development and the adoption of high-yielding cultivars have been crucial in eliminating hunger. The green revolution technologies of the past, on the other hand, were concentrated on a restricted number of cereal crops and excelled with favorable environmental conditions and high input regimes. This has resulted in decreased agricultural and food diversity and in certain cases, environmental issues linked to the overuse of agrochemical inputs.

Plant breeding contains genetically modified organisms (GMOs), and crops that have been genetically modified can assist to further boost yields while resolving the drawbacks of green revolution technology. To name a few of the technologies on which plant biotechnologists have already focused heavily, new plant breeding technologies can help to increase crop diversity, raise yield potentials, provide better resistance to abiotic stresses, pests, and diseases, and boost nutrient use efficiency, make crops more resilient to extreme temperatures, and improve nutritional value. A few GMO crops have already gained widespread adoption and demonstrated positive effects on the economy, society, and the environment, even in the small farm sector of developing nations.

References

Borlaug, N.E. "Contributions of Conventional Plant Breeding to Food Production." Science 219 (1983):689-93.

Brooks, H.J., Vest G. "Public Programs on Genetics and Breeding of Horticultural Crops in the United States." HortScience 20 (1985):826-30.

Ceccarelli, S., Grando S. 1993. "From Conventional Plant Breeding to Molecular Biology." International Crop Science I. Edited by Buxton D.R., Shibles R., Forsberg R.A., Blad B.L., Asay K.H., Paulsen G.M., et al. Madison: Crop Science Society of America, Inc. pp. 533-37.

Duvick, D.N. 1986. "Plant Breeding: Past Achievements and Expectations for the Future." Economic Botany 40 (1986):289-97.

Fehr, W.R. 1984. Genetic Contributions to Yield Gains of Five Major Crop Plants. Madison: Crop Science Society of America, Inc.

Food and Agriculture Organization of the United Nations. 2002. FAOSTAT Agriculture Data. Rome: Food and Agriculture Organization of the United Nations.

Hartung, F., Schiemann J. "Precise Plant Breeding Using New Genome Editing Techniques: Opportunities, Safety and Regulation in the EU." The Plant Journal 78, 5 (2014):742-52.

Hayes, H.K. "A Half-century of Crop Breeding Research." Agronomy Journal 49 (1957): 626-33.

Jain, S. 1992. Crops and Man, 2nd edition. Edited by Harlan J.R. Madison: American Society of Agronomy.

Johnson V.A. "What Makes a Successful Plant Breeder?" Journal of Agronomic Education 10 (1981):85-6.

Poehlman, J.M., Quick J.S. 1983. "Crop breeding in a hungry world." Crop Breeding, edited by D.R. Wood. Madison: American Society of Agronomy and Crop Science Society of America. pp. 1-19.

Roll-Hansen, N. 1986. "Svalöf and the Origins of Classical Genetics." Svalöf 1886-1986, Research and Results in Plant Breeding. Edited by Olsson G. Stockholm: LTs Förlag. pp. 35-43.

Singh, D.P., Singh, A.K., Singh, A. 2021. "Plant Breeding: Past, Present, and Future Perspectives." Plant Breeding and Cultivar Development. Cambridge: Academic Press. pp. 1-24.

2

Modern and Transgenic Breeding Approaches

Vignesh S.

Department of Biotechnology, Centre for Plant Molecular Biology and Biotechnology Tamil Nadu Agricultural University, Coimbatore, Tamil Nadu, India

Abstract

Modern and transgenic breeding approaches consist of modern plant-breeding triangle, high-throughput phenotyping (HTP), marker-assisted selection (MAS), marker-assisted backcrossing (MABC), genomic selection (GS), genome-wide association study (GWAS), and genome editing. MAS accelerates trait identification through DNA markers, improving efficiency in selecting plants with desired traits. MABC employs molecular markers to facilitate gene introgression from donor parents (DPs) to superior recurrent parents (RPs). GS employs genome-wide markers to predict performance, revolutionizing breeding by considering both major and minor-effect quantitative trait loci (QTLs). Genome editing techniques like CRISPR-Cas9 (clustered regularly interspaced short palindromic repeats–associated protein 9) enable targeted gene modifications for desired traits. Genome editing involves selecting target genes, designing guide RNAs for Cas9, and introducing the CRISPR-Cas9 system into plants. This technology has been utilized to enhance crop productivity, sustainability, and disease resistance. Transgenic crops are designed to carry specific traits, such as pest resistance, virus resistance, abiotic stress tolerance, and nutritional enrichment. Agrobacterium tumefaciens is highlighted as a tool for transgene introduction, offering advantages like low copy number integration and stable expression. Genetic engineering remains a promising avenue for creating disease-resistant crops and ensuring global food security.

Keywords: *Modern plant-breeding triangle, marker-assisted breeding, genomic selection, genome-wide association study, genome editing*

Introduction

Plant breeding has historically played a pivotal role in transforming wild plants into cultivated crops, adapting their genetics to align with changing human and environmental demands. Innovations such as hybrid varieties and semidwarf cereal genotypes drove remarkable increases in crop productivity, famously exemplified by the "green revolution" in nations like India. However, with the global population on track to reach nine billion by 2042 and climate-induced stresses reshaping agricultural landscapes, breeders now face the formidable challenge of rapidly developing high-yield, stress-tolerant crop genotypes.

Traditional breeding methods, rooted in phenotype-based selection, present limitations due to their unpredictability, high costs, and inability to forecast progeny performance from phenotypic traits accurately alone. To overcome these constraints, DNA markers like simple sequence repeats (SSRs) and single-nucleotide polymorphism (SNPs) have been devised, enabling more precise, efficient selection. These markers facilitate gene introgression, recurrent selection, diversity analysis, germplasm characterization, and the determination of genetic purity in hybrid seed lots. As markers progressively assume a central role in modern breeding programs, underscoring the need for comprehensive resources detailing marker integration in plant breeding strategies.

Genetic engineering offers a potent solution to the challenges of conventional breeding. Through techniques like CRISPR-Cas9 (clustered regularly interspaced short palindromic repeats–associated protein 9), desired genes can be directly transferred across species to confer specific agronomic traits. This approach has yielded successes such as antibiotic-resistant tobacco plants, herbicide-resistant crops, and insect-resistant varieties. While some countries withdrew early transgenic crops, many have embraced biotech solutions to address global issues like hunger, malnutrition, and climate change. Notably, herbicide resistance has emerged as a widely employed trait, with glyphosate-resistant crops taking the forefront. These advancements underscore the critical role of genetic engineering in increasing agricultural productivity and addressing pressing global challenges.

Modern Breeding Approaches

Modern technologies have merged with conventional breeding to expand plant breeding goals beyond yield improvement, targeting traits like nutrition, and environmental responsiveness. Strategies involve marker-assisted selection (MAS), quantitative trait loci (QTL) mapping, gene editing, and genomic selection (GS) for improved selection efficiency. Modern breeding relies on stable genotypes, overcoming genotypes by environment interactions for effective results. Overall, integration of scientific research into breeding makes modern plant breeding more reliable and efficient.

Modern Plant-breeding Triangle

The evolution of plant breeding from traditional to modern methodologies highlights the integration of genomics, phenomics, and enviromics to enhance genetic gain. Conventional breeding programs have shown increasing genetic gain, but the emergence of modern technologies presents opportunities for further improvement (Fig. 1).

Genomics Training and Prediction.
Prediction of Complex Trait of Testing Population for Genomics
GxE Models Including Multi-trait, Multi-environment Data

Environmics Climatic and Soil Data.
Structure Environmental Data
for Explaining Causes of GxE

Phenomics High Throughput Phenotype (HTP).
Image Analyses to Predict Visual
Assessment of Different Traits

Fig. 1: Modern plant breeding integrates genomics, phenomics, and enviromics for genetic gains. (G × E: Gene–environment)

The challenges lie in effectively integrating these technologies, as evidenced in rice breeding. Genomics rapid cycling, high-throughput phenotyping (HTP), and enviromics are identified as key technologies with the potential to enhance conventional breeding and boost genetic gain. The chapter envisions a more interconnected approach to plant breeding, leveraging genomics, phenomics, and enviromics to develop resilient cultivars capable of addressing future environmental challenges due to climate change. This integration is expected to create a comprehensive strategy that enables informed decision-making and the development of robust crop varieties.

High-throughput Phenotyping

Challenges like resource scarcity, climate change, and population growth necessitate sustainable enhancements in crop yield and quality. Genetic advancement through breeding is vital for improved crop productivity. Functional genomics has enabled genome sequencing and identification of genes affecting key traits, yet underutilization of genomic data due to limited crop phenotypic information remains a constraint. Efficient technologies capturing phenotype-genome links are crucial for holistic crop improvement.

Plant phenomics involves multidimensional phenotype acquisition across developmental levels. Phenotypic performance results from genotype–environment interactions. HTP platforms, utilizing advanced sensors and imaging, are categorized as microscopic, ground-based, and aerial systems, aiding trait analysis at varying scales.

Microscopic phenotyping examines detailed traits in plant tissues using techniques like X-ray micro-CT (computed tomography) for nondestructive analysis of traits such as kernel attributes, stalk anatomy, and root vascular bundles. Ground-based phenotyping includes portable, stationary, and movable platforms with sensors for estimating traits. Deep learning, a subdivision of machine learning, has automated trait extraction using convolutional neural networks (CNNs), aiding the identification, counting, and prediction of crop traits.

Crop models, including functional structures, simulate growth and yield, enhanced by remote sensing and geographic systems. In conclusion, efficient phenotyping and data analysis, driven by imaging and deep learning, are pivotal for tackling agricultural challenges, guiding breeding, resource management, and yield projection.

Marker-assisted Selection

Conventional plant breeding relies on selecting desirable plants influenced by genotype-environment interactions. Cross-breeding plants with desired traits and successive selection over generations under varying conditions is a fundamental approach. Breeders work to integrate both qualitative and quantitative traits from donor plants into elite cultivars. Introducing single-gene controlled traits enhances trait expression, often achieved through backcross (BC) breeding for qualitative traits like disease resistance, despite challenges of linkage drag and time-consuming processes.

DNA markers are reliable tools for selection, offering stability and insensitivity to environmental conditions. They surpass phenotypic assays in various ways, increasing reliability by minimizing errors from environmental factors and gene interactions. Additionally, they streamline efficiency by enabling seedling-stage scoring, especially for traits expressed later in development, reducing time and space requirements. DNA markers also lower costs compared to growing plants for phenotypic assessment. Before adopting DNA marker-assisted methods, breeding program leaders must evaluate practical considerations and costs, particularly beneficial for complex quantitative traits with low heritability. Developing marker assays for such traits may be intricate and costly, but the knowledge gained facilitates effective marker-assisted breeding, reducing reliance on phenotypic assays.

Molecular marker technology introduces MAS as a way to accelerate and enhance breeding. MAS involves selecting plants based on marker genotypes, expediting trait identification before field evaluation. MAS encompasses methods like marker-assisted introgression, population screening, and gene pyramiding, along with more advanced techniques like marker-based recurrent selection and molecular-phenotypic selection. MAS holds potential to improve breeding efficiency and shorten development time, particularly beneficial due to the intricate nature of genotype-environment interactions.

Marker-assisted selection is a potent tool in plant breeding, effectively integrating target genes into populations for enhanced efficiency. MAS complements conventional methods, aiding in the incorporation of major genes and QTL as well as stacking resistance genes for broader protection against diseases and pests. Its advantages include early seedling-stage selection, independence from environmental effects, identification of recessive alleles in heterozygotes, support for gene pyramiding, and improved progress for low-heritability traits. MAS eliminates unreliable phenotypic evaluations, suits traits with impractical assessments, offers cost-effectiveness, allows multi-marker assessment from a single DNA sample, and assists in parent selection. It accelerates breeding, addresses economic constraints, and serves diverse purposes, making it an indispensable tool for modern plant breeders seeking enhanced selection efficiency and streamlined development.

Marker-assisted Backcrossing

Backcross breeding involves transferring a desired trait from a donor parent (DP) to a recurrent parent (RP), a superior variety lacking the trait. The process starts with an F1 cross between DP and RP, followed by backcrossing to the RP for several generations. This gradually replaces the DP genome with the RP genome, resulting in around 1.6% of DP genome after five backcrosses. Phenotypic selection for RP traits further enhances this effect. After sufficient backcrosses, progeny plants similar to RP but carrying the target trait are selected. This process, known as gene introgression, aims to improve existing varieties and is particularly useful when introducing transgenes.

Foreground Selection

Foreground selection (FG), proposed by Tanksley (1983) and named by Hospital and Charcosset (1997), involves selecting target genes based on linked markers, advantageous for challenging phenotypic assessments. Useful for transferring multiple QTLs or traits, it enhances resistance combinations. Proximity between marker and target gene/QTL impacts effectiveness; <5 cM proximity is ideal. Flanking markers on either side of the gene/QTL improves accuracy. MAS in plant breeding integrates target genes efficiently, combining conventional methods. It is valuable for major gene introgression, pyramiding resistances, and traits with low heritability. MAS's benefits include early seedling-stage selection, environmental insensitivity, identifying recessive alleles, gene stacking, and cost-effective evaluation. It aids in parent selection, cultivar assessment, and genetic studies, enhancing breeding efficiency and addressing constraints (Fig. 2A).

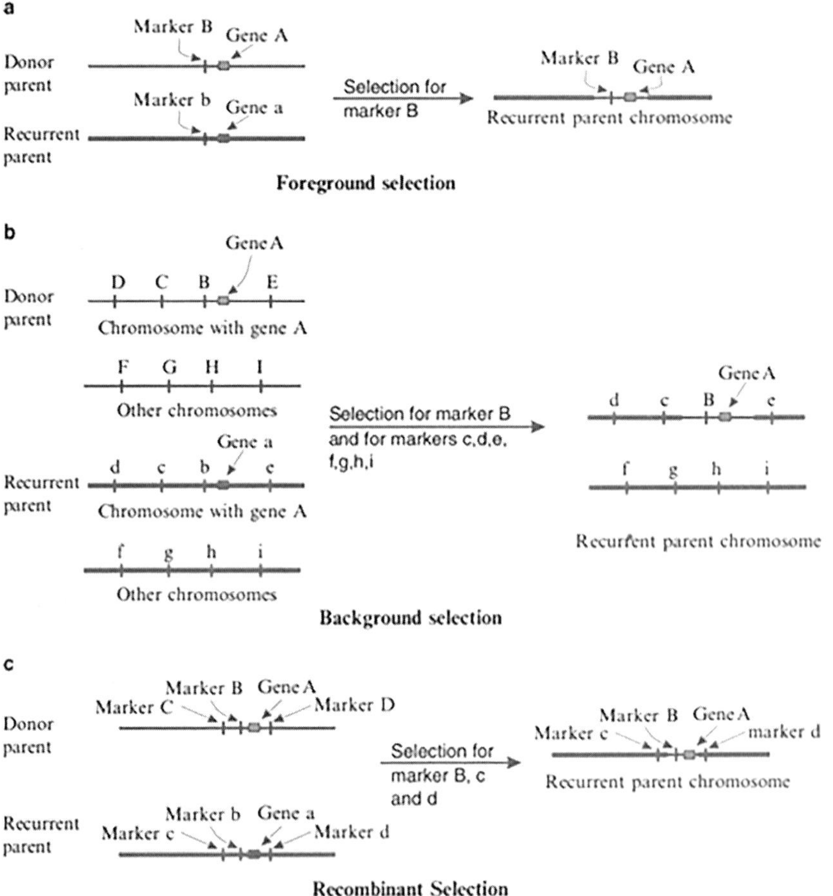

Figs. 2 A to C: *A,* Foreground selection targets the target gene using tightly linked markers. B, Background selection focuses on recovering the recurrent parent genotype using genome-wide markers. C, Recombinant selection eliminates donor genome flanking the target gene, selecting for nearby recurrent parent markers.

Recombinant Selection

Foreground selection is used in plant breeding for both simple and complex traits. But for complex traits, such as quantitative traits, recombinant marker selection is employed, utilizing both peak and flanking markers. Recombinant marker selection minimizes transferred chromosomal regions and prevents linkage drag. FG is conducted in the first BC generation using tightly linked peak markers. Plants positive in FG are tested for flanking markers to detect recombination events. This method reduces undesirable linked genes during gene transfers. It aims to transfer the target gene/QTL with the minimal undesired genome (DP) using markers located within ~1 cM of the target gene, minimizing linkage drag. Two strategies exist single-generation selection, requiring extensive progeny screening and a more effective two-generation approach that reduces screening numbers, making the process feasible in breeding programs (Fig. 2C).

Background Selection

Marker-assisted background selection accelerates the recovery of the RP genome in plant breeding by employing unlinked markers to the target gene/QTL. Introduced by Young and Tanksley and termed by Hospital and Charcosset, this method reduces the need for six conventional backcross generations to as early as BC2, aided by flanking markers to minimize linkage drag. Its effectiveness depends on marker-to-gene distance, trait heritability, and gene quantity. When combined with phenotypic selection, it optimizes resource utilization and aids gene stacking. Success relies on marker density and the RP's genetic background, thereby enhancing gene pyramiding.

Molecular markers aid in achieving objectives: Markers linked to the target gene enable direct selection (FG), codominant markers across the genome select for high RP genome proportion (background selection), and markers flanking the gene enable selection for rare recombinants lacking donor genome (recombinant selection). This approach called marker-assisted backcrossing (MABC), is crucial for efficiently transferring traits while preserving desired genome proportions. MABC is used for foreground, background, and sometimes recombinant selection, and it is particularly valuable for improving existing varieties with specific traits (Fig. 2B).

Genomic Selection

Marker-assisted selection is vital for traits influenced by oligogenes and significant QTLs. It is used in backcross breeding and marker-assisted recurrent selection (MARS), particularly for stacking beneficial genes/QTLs. Backcross breeding improves a parent's traits through introgressed genes/QTLs but does not create new gene combinations. While effective for major-effect

QTLs, MAS does not suit quantitative traits linked to minor-effect QTLs. MARS critiqued for relying on markers associated with traits counters this by combining marker and trait data into a selection index to address small-effect QTLs.

Genomic selection, introduced to improve MAS and MARS, utilizes genome-wide marker data regardless of trait associations. Proposed by Meuwissen *et al.* in 2001, GS addresses prior limitations by acknowledging traits' mixed QTL nature, some with large and some with small effects. By using comprehensive marker datasets, GS enhances selection accuracy for both major and minor-effect QTLs, revolutionizing breeding and contributing to successful commercial varieties.

Genomic selection selects complex traits via advanced marker technology and statistics, covering technology, genetics, markers, accuracy, nonadditive effects, subpopulation impact, and long-term gains. Requiring markers proportional to genetic map length and population size, GS accelerates breeding while preserving diversity, pivotal for future plant breeding due to cost reductions and statistical advancements.

Genomic selection involves a training population for model refinement and a breeding population for improvement. Training refines the model, providing marker-associated effects for genetic estimated breeding values (GEBVs). Breeding uses GS to select superior lines for new varieties (Fig. 3).

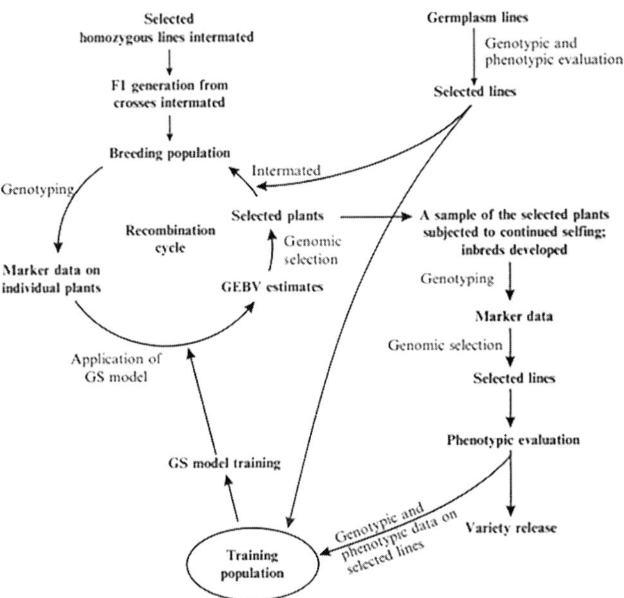

Fig. 3: A recurrent selection scheme for genomic selection (GS) in a self-pollinated crop involving new germplasm, pure line development, recombination cycles, and constant genetic estimated breeding values (GEBV) accuracy maintenance.

Steps include training, genotyping, phenotyping, computation, GEBV calculation, and selecting lines based on GEBVs. With a suitable strategy, GS achieves breeding goals. Long-term selection benefits from diverse training sets and preserving rare alleles. Limited crop accuracy data require empirical studies. Cross-validation's limitations due to generational gaps should be noted. Public breeding programs find promise in GS. Optimizing breeding cycles and possibly modeling epistasis enhance GS impact without exhaustive biological knowledge.

Genomic-wide Association Selection

Genomic selection helps to predict performance using markers across the genome, while the genome-wide association study (GWAS) helps to identify specific genetic markers associated with a trait of interest. Both of them have their applications, and they can complement each other in molecular breeding programs.

Genome-wide association study holds immense significance in both plant and animal breeding, serving as a cornerstone for revealing the intricate connections between genotypes and phenotypes through rigorous statistical methodologies.

The pivotal role of GWAS lies in its ability to identify specific genomic regions closely tied to traits, thus fostering advancements in yield enhancement and adaptability. Genetic markers unveiled by GWAS play a crucial role in breeding by enabling the prediction of expected breeding values (EBVs), thereby facilitating the selection of superior candidates for further breeding.

Genome-wide association study employs linear regression and least squares techniques to unravel the intricate interplay between genotypes and phenotypes. This methodology is especially advantageous for traits influenced by genes with notable effects. In contrast, Genomic Prediction (GP) is centered around predicting forthcoming phenotypic outcomes by computing individual genetic values, thus streamlining the selection process for potential candidates. GP harnesses mixed linear models to collectively utilize all markers, enhancing its predictive accuracy.

The GWAS offers insights into the identification of QTLs, some of which hold sway over gene structure or expression. Meanwhile, the ability of GP to predict the cumulative effects of multiple genes bolsters the comprehension of evolutionary dynamics, encompassing forces such as selection, introgression, and inbreeding.

Transgenic Breeding Approaches

The development of transgenic crops since the 1990s marks a significant milestone in crop improvement history. While transgenic crops have grown in area and spread across nations, concerns about biosafety, gene movement, and environmental impact persist. Crop choice for transgenic improvement should prioritize national importance, nutritional security, and economic benefits. Traits controlled by major genes and those not easily improved conventionally are suitable for transgenic engineering. Pest and virus resistance, abiotic stress tolerance, nutritional enrichment, and shelf life extension are key target traits. Genes sourced from various organisms allow for novel trait engineering. Transgenic crops should align with national agricultural goals, ensuring long-term benefits and minimal biosafety risks.

Agrobacterium Tumefaciens

Agrobacterium tumefaciens, a plant pathogen, is widely used to introduce foreign genes into plants, creating transgenic organisms. This bacterium naturally infects wounds in dicotyledonous plants, causing crown gall disease. Its tumor-inducing (Ti) plasmid triggers tumor formation by expressing T-DNA genes, which are stably integrated into the plant genome. Optimizing techniques, like preculturing explants and modifying conditions, enhances transformation efficiency. Genetic engineering enables genes from various organisms to be transferred, yielding genetically modified organisms with desirable traits. *Agrobacterium*-mediated transformation, utilizing *A. tumefaciens*, is advantageous due to its broad crop range, low copy number integration, and stable expression. However, achieving high efficiency remains challenging. This method involves integrating DNA segments into plant chromosomes and regenerating transgenic plants from transformed cells. Despite the technique's benefits, achieving high transformation efficiency and gene expression is still a challenge.

Genome Editing

The challenges posed by diminishing farmland, water resources, and climate change necessitate advancements in genetic modification (GM) and breeding for food production. Early human selection of desirable traits in plants before understanding genetics, such as maize from teosinte, illustrates a historic form of GM. To accelerate plant breeding, genetic engineering techniques like genome editing using programmable endonucleases have emerged. These techniques induce targeted breaks in genes, allowing DNA repair and modifications. Three main programmable endonucleases—zinc finger nucleases, TALENs (transcription activator-like effector nucleases), and CRISPR-Cas9—are used for plant genome editing.

CRISPR-Cas9, involving Cas9 nuclease and guide RNA, stands out for its simplicity, efficiency, and versatility. It finds applications in medicine and agriculture. While CRISPR-Cas9 has been utilized to treat human diseases and enhance virus resistance, its potential for crop improvement is especially noteworthy which focuses on CRISPR-Cas9 applications in boosting crop productivity and sustainability.

The process of genome editing involves identifying target genes for desired traits, designing guide RNAs to direct Cas9, and utilizing tools for minimal off-target effects. Dynamic expression vectors facilitate co-expression of Cas9 and guide RNAs in plant cells. Plant transformation follows a procedure akin to transgenic plant generation. The vector carrying Cas9 and guide RNA is introduced into *A. tumefaciens* or *Rhizobium rhizogenes* for plant transformation. Transgenic plants expressing the CRISPR-Cas9 cassette are identified through selection markers, followed by sequencing to detect genome editing-induced mutations. In sexually propagated plants, Mendelian segregation can eliminate the Cas9 transgene in subsequent generations, yielding edited plants without transgenes. Protocols have been developed for transgene-free genome editing using guide RNA-Cas9 ribonucleoprotein complexes or transient expression. In species resistant to *Agrobacterium* transformation, *R. rhizogenes*-mediated transformation offers a solution. This approach has been applied successfully to plants like soybean, tomato, and rubber dandelion.

Insect-resistant Crops by Transgenic Breeding Approaches

Insect-borne pathogens are increasingly impacting health, environment, and agriculture, spurring interest in control strategies. The approach involves planting insect-resistant seeds, like rice TPS46, which enhances crop protection. The TPS46 gene contributes to volatile production and innate immunity in rice; its silencing renders the plant more vulnerable. Insect-resistant crops having *Bacillus thuringiensis* (Bt) genes have transformed pest management. Notably, Bt cotton, introduced in 1996, effectively combats bollworms, reducing pesticide use and environmental harm. GM methods, such as Bt technology, offer a pathway for sustainable pest control, delivering substantial yield and economic advantages.

The current absence of insect–pest resistance in many crops has led to extensive use of hazardous chemical controls, detrimental to consumers and the environment. Genetic improvements minimizing chemical applications offer significant benefits for growers. The Bt (Cry) gene sourced from *B. thuringiensis*, isolated from soil bacteria, has demonstrated efficacy against various lepidopteran insects across multiple crops. Insect resistance via GMs was first achieved in 1987, conferring resistance in transgenic Bt tomato plants against specific pests. This approach has expanded to various crops, including fruit trees, brinjal (egg plant), potato, cabbage, okra, and kiwifruit, introducing genes for insecticidal crystal proteins and resistance factors. These modifications, often driven by specific promoters, show promise in reducing reliance on harmful chemicals, promoting sustainable pest management, and fostering healthier crop production.

Disease Management

Genetic engineering and genome editing offer efficient alternatives to traditional plant breeding for enhancing crop disease resistance. Slow classical breeding methods have led to exploring CRISPR-Cas9 and other precise genome editing techniques. Understanding plant disease mechanisms aids sophisticated approaches to boosting resistance. Utilizing disease-resistant genes from wild relatives shows potential for sustainable control, despite linkage drag challenges. These advancements promise improved agricultural productivity and food security by creating disease-resistant crops.

RNA-based genetic engineering has facilitated transgenic crop development for viral disease resistance, including against papaya ringspot, potato virus, and maize viruses. Pathogen adaptability drives continuous innovation.

CRISPR-Cas9, for instance, targets genes to engineer late blight resistance in potatoes. Fungal pathogens imperil crops, and CRISPR-Cas9 displays promise in deactivating genes like *DMR6* to confer resistance. Explorations include combining resistant genes and transferring immunity genes to manage fungal diseases.

Genetic engineering addresses challenges posed by bacterial diseases like tomato bacterial spot through techniques like transgenic tomato plants expressing the *Bs2* gene, resulting in heightened disease resistance and yield. Merging single-resistance genes with gene inactivation methods like *DMR6* offers strategies for broad, lasting disease resistance. Despite progress, genetic diversity, new pathogen strains, and public acceptance remain concerns. Genetic engineering holds potential to create disease-resistant crops, fostering sustainable agriculture and global food security. Approaches involving systemic acquired resistance (SAR), anti microbial proteins like chitinase and defensin as well as RNA interference (RNAi), have shown success against various pathogens, enhancing crop resilience and productivity.

Conclusion

Plant domestication shaped natural genetic diversity through human selection, yielding diverse plant phenotypes. Traditional farming further enriched genetic diversity. Plant domestication and traditional farming have shaped genetic diversity, but modern demands exceed their pace. Scientific breeding, aided by molecular tools, accelerates varietal improvement. Expectations for addressing contemporary challenges are high using ancestral genetic resources. CRISPR/Cas advances enable precise genome editing, yet a functional understanding of genetic networks is crucial. Challenges like delivery methods and gene targeting persist. MAS enhances breeding through increased gain and stacked alleles. Private sectors adopt MAS, but the public sector faces obstacles like limited resources and training. Genomics, reduced marker costs, and virtual platforms could ease adoption. Integrating disciplines and genomics research aids metabolic pathways understanding. High-throughput approaches, automation, and bioinformatics bolster crop improvement, but will not replace conventional methods. A genotype's value is determined by performance and acceptance in the target environment.

References

Alemayehu, D. "Application of Genetic Engineering in Plant Breeding for Biotic Stress Resistance." International Journal of Research Studies in Biosciences (IJRSB) 5, 9(2017):28-35.

Anwar, A., Kim J. "Transgenic Breeding Approaches for Improving Abiotic Stress Tolerance: Recent Progress and Future Perspectives." International Journal of Molecular Sciences 21, 8 (2020): 2695.

Bhat, S. 2005. "Transgenic Crops: Priorities and Strategies for India." Current Science 88, 6(2005):770-2.

Boopathi, N.M. 2020. Genetic Mapping and Marker Assisted Selection: Basics, Practice and Benefits, 2nd ed. Singapore: Springer Nature.

Crossa, J., Fritsche-Neto R., Montesinos-Lopez O.A., Costa-Neto G., Dreisigacker S., Montesinos-Lopez A., et al. "The Modern Plant Breeding Triangle: Optimizing the Use of Genomics, Phenomics, and Enviromics Data." Frontiers in Plant Science 12 (2021).

El-Mounadi, K., Morales-Floriano M.L., Garcia-Ruiz H. "Principles, Applications, and Biosafety of Plant Genome Editing Using CRISPR-Cas9." Frontiers in Plant Science 11 (2020).

Gao, C. "Genome engineering for crop improvement and future agriculture." Cell 184, 6(2021):1621-35.

Hua, K., Zhang J., Botella J.R., Ma C., Kong F., Liu B., et al. "Perspectives on the Application of Genome-Editing Technologies in Crop Breeding." Molecular Plant 12, 8(2019):1047-59.

Jannink, J.L., Lorenz A.J., Iwata H. "Genomic Selection in Plant Breeding: From Theory to Practice." Briefings in Functional Genomics 9, 2 (2010):166-77.

Lamichhane, S., Thapa S. "Advances from Conventional to Modern Plant Breeding Methodologies." Plant Breeding and Biotechnology 10, 1 (2022):1-14.

Lorenz, A.J., Chao S., Asoro F.G., Heffner E.L., Hayashi T., Iwata H., et al. 2011. "Genomic Selection in Plant Breeding: Knowledge and Prospects." Advances in Agronomy, vol. 110. Amsterdam: Elsevier Inc.

Parmar, N., Singh K.H., Sharma D., Singh L., Kumar P., Nanjundan J., et al. "Genetic Engineering Strategies for Biotic and Abiotic Stress Tolerance and Quality Enhancement in Horticultural Crops: A Comprehensive Review." 3 Biotech 7, 4(2017):239.

Singh, B.D., Singh A.K. 2015. Marker-Assisted Plant Breeding: Principles and Practices. New Delhi: Springer.

Song, P., Wang J., Guo X., Yang W., Zhao, C. "High-throughput Phenotyping: Breaking through the Bottleneck in Future Crop Breeding." The Crop Journal 9, 6(December 2021):633-45.

Yildiz, M., Aycan M., Park S. 2016. "New Approaches to Agrobacterium tumefaciens-Mediated Gene Transfer to Plants." Genetic Engineering - An Insight into the Strategies and Applications. InTech. DOI:10.5772/66465.

Zhang, Q., Zhang Q., Jensen J.J. "Association Studies and Genomic Prediction for Genetic Improvements in Agriculture." Frontiers in Plant Science 13(2022).

3

Breeding Strategies for Disease Resistance

***Himanshi Jhalora*[1], *Radhika Shekhawat*[2] *and Parul Gupta*[3]**

[1–3]*Department of Genetics and Plant Breeding, Maharana Pratap University of Agriculture and Technology, Udaipur, Rajasthan, India*

Abstract

The chapter "Breeding Strategies for Disease Resistance" provides a comprehensive exploration of addressing disease resistance through various breeding strategies. In an era of escalating global challenges and climate change, where crop failures can lead to food shortages and economic instability, the chapter delves into strategies that integrate traditional breeding methods with cutting-edge genetic advancements to strengthen plants against pathogens. Understanding the genetics of plant–pathogen interactions holds profound importance. This knowledge elucidates the mechanisms underpinning disease development and offers insights into developing targeted strategies for disease management. By deciphering the genetic details governing host susceptibility and pathogen virulence, we gain the ability to engineer crops with enhanced resistance, reducing reliance on chemical interventions and promoting sustainable agriculture. The chapter further discusses the mechanism of disease resistance in plants which involves inherent genetic traits that identify pathogen molecules, triggering a rapid defense response called "gene-for-gene" interaction. Systemic acquired resistance (SAR), where a local pathogen attack prompts the entire plant to boost its defense preparedness. Signaling molecules like salicylic acid play a pivotal role. Additionally, plants employ chemical defenses via secondary metabolites and leverage beneficial microorganisms to stimulate immune pathways. This coordinated approach empowers plants to effectively counter pathogens and elevate disease resistance. In this chapter, we discuss breeding strategies for disease resistance that encompass a range of vital approaches. Traditional methods involve selecting and breeding plants with inherent resistance traits and incorporating genes from related species to enhance defenses. Modern techniques like marker-assisted selection, genomic selection, and gene editing expedite targeted trait acquisition.

Moreover, gene pyramiding combines multiple resistance genes for enhanced protection. These strategies, merging traditional understanding with pioneering technology, strengthen plants to resist diseases ensuring sustainable agricultural practices and global food security.

Keywords: *Breeding strategy, climate change, disease resistance, gene pyramiding, marker-assisted selection*

Introduction

In an ever-changing world where agriculture plays a very important role in feeding the growing population, the threat of plant diseases commands considerable attention. Disease outbreaks result in devastating losses of crops, impacting food security and livelihoods. Breeding strategies aimed at enhancing disease resistance in plants have become a vital focus for researchers and plant breeders alike.

Disease is a deviation or disruption from the normal functioning or structure of an organism, often resulting in physiological, biochemical, or anatomical abnormalities. Plant disease refers to the physiological and structural abnormalities caused by a variety of pathogens, including fungi, bacteria, viruses, nematodes, and other microorganisms, that negatively impact the growth, development, and overall health of plants. These diseases can result in symptoms such as wilting, discoloration, stunted growth, tissue necrosis, and reduced yield. Plant diseases can have far-reaching consequences, affecting agricultural productivity, ecosystem dynamics, and food security (Agrios 2005). The plant affected by the disease is known as *the host*, and the organism that produces the disease is known as the *pathogen*. This chapter examines the various approaches and techniques used to develop plants with heightened disease resistance, ensuring a more sustainable and secure agricultural future (Mehrotra 2017).

Losses due to Diseases

Losses due to plant diseases can have significant economic, social, and environmental impacts on agriculture and food production systems. These losses can manifest in several ways:

- ***Yield reduction***: Disease-affected plants will result in general stunting, the killing of branches and other parts of plants, damage to leaf tissues, and damage to reproductive organs like fruits and seeds and overall reducing the yield.
- ***Quality decline***: Diseases can compromise the quality of harvested crops by affecting their appearance, taste, texture, and nutritional content.

- ***Environmental impact***: Some plant diseases can lead to defoliation, dieback, or death of plants, affecting natural ecosystems and biodiversity. This can alter ecosystem dynamics and disrupt the balance of local flora and fauna.
- ***Loss of genetic diversity***: Disease outbreaks can lead to the loss of plant varieties that are susceptible to the disease, reducing genetic diversity within crop populations. This can weaken the resilience of crops to future disease threats. The effect of diseases may lead to the disappearance of the cultivation of otherwise superior but susceptible varieties. One prominent example is the Gros Michel banana variety, which faced near-extinction due to a devastating fungal disease called Panama disease which is caused by *Fusarium oxysporum f. sp. cubense.*

Diseases reduce biomass and consequently, yield. Often the losses may range from 20% to 30%, and in cases of severe infection, the total crop may be lost.

Breeding strategies for disease resistance in plants involve the selection and breeding of plants to develop varieties that exhibit improved resistance to various pathogens. These strategies aim to enhance the natural defense mechanisms of plants against diseases, reducing the need for chemical interventions and promoting sustainable agriculture. The process typically involves both conventional breeding techniques and advanced genetic technologies. Later in this chapter, you will find a discussion of some important breeding strategies for disease resistance.

Genetics of Plant–Pathogen Interaction

Each host–parasite interaction is a struggle for survival between two organisms. In nature, there exists a state of balance, and surviving plants and their parasites/pathogens are capable of coexistence. This discussion focuses on deciphering how plants perceive and respond to pathogenic threats at a molecular level as well as how pathogens evolve strategies to manipulate host defenses. The genetics of plant–pathogen interactions explores the intricate genetic mechanisms governing the dynamic relationship between plants and pathogens. This field focuses on deciphering how plants perceive and respond to pathogenic threats at a molecular level as well as how pathogens evolve strategies to manipulate host defenses.

Nature of Resistance

Host resistance may be true resistance that operates when the host and the pathogen come into contact with each other or it may be due to mechanisms, collectively called "avoidance". Avoidance reduces the contact between the host and the pathogens. Examples of avoidance are (1) resistance of barley cultivars with closed flowers to *Ustilago nuda*; (2) resistance of tall wheat cultivars to glume blotch, *Septoria nodorum* (the height of the plant reduces the spread of inoculum to the upper parts of the plant); (3) resistance of rubber variety LCB870 to mildew and coffee cultivars SRG and Dilla to coffee berry disease due to reduced period of susceptibility; and (4) reduced susceptibility of the pole and indeterminate types of bean cultivars, compared to bush-type cultivars, and to web blight (caused by *Rhizoctonia solani*). The mechanisms of avoidance or escape, if identified, can be used as valuable characteristics in designing breeding programs (Hooker 1980).

Mechanism of Disease Resistance

Different mechanisms operate in plants for disease resistance; *viz.*, (1) resistance to the establishment of pathogen in the host tissue; (2) resistance to the growth and development of the pathogen already established in the host tissue; and (3) ability of host to perform well despite the establishment of the pathogen in the host tissues. The first two mechanisms are considered as true resistance and the last is tolerance.

True resistance: Vertical resistance is associated with hypersensitivity of the host cells leading to the death or chlorosis around the point of infection and consequent starving of the obligate parasite.

Hypersensitivity in disease resistance refers to a rapid and localized defense response exhibited by plants when they encounter pathogenic microorganisms, such as bacteria, viruses, or fungi. This response involves the programmed cell death of infected plant cells at the site of infection, effectively limiting the spread of the pathogen. Hypersensitivity is a critical mechanism that helps plants contain and control infections, preventing the pathogen from spreading throughout the entire plant. This process involves various signaling pathways and molecular components that activate defense mechanisms, ultimately enhancing the plant's resistance to diseases.

Disease resistance is governed by several morphological, physiological, and biochemical features of the host plant. *Morphological* features of hosts for disease resistance (such as small, few sunken, and hairy stomata with lesser opening duration of flower) refer to physical structures or traits that plants evolved to deter or prevent pathogen infection. These features act as barriers that hinder the establishment and spread of pathogens. *Physiological* characteristics like high osmotic pressure and acidity of cell sap and high crude fiber content have been reported to confer disease resistance. *Biochemical* studies have revealed that high contents of tannin, protocatechuic acid, and catechol, phenol, alkaloids, silica, and riboflavin were associated with disease resistance in different crop plants. Most of these chemical compounds are toxic to pathogens, and thus check the establishment of parasites into the host tissue. *Nutritional* feature constitutes the reduction in growth and spore production of pathogen due to unfavorable physiological conditions within the host. Moreover, a resistant host will limit the growth and reproduction of the pathogen by compromising the nutritional requirements of the pathogen.

Tolerance: The ability of a host to reproduce well despite the establishment of pathogen in the host tissue is referred to as tolerance. Unlike other mechanisms like hypersensitivity or resistance, where the goal is to prevent or limit pathogen proliferation, tolerance focuses on reducing the harm caused by the pathogen's presence. The performance of tolerant plant is generally similar to resistant plants. However, such plants fail to check the entry and establishment of a pathogen in the host cell. Tolerant plants are always put in the category of susceptible plants because they exhibit the symptoms of disease infestation (Singh 2018).

Disease Escape

As implied by its name, disease escape pertains to the absence of disease incidence in susceptible host plants or varieties solely due to environmental factors. Disease escape predominantly transpires through the evasion of pathogen contact, facilitated by adverse weather conditions that hinder infection. However, within the context of disease resistance breeding, disease escape poses a challenge to the identification of genuinely resistant plants. This challenge arises because the absence of disease in susceptible plants could be misconstrued as resistance. Disease escape is attainable through (1) early varieties, (2) altered date and site of planting, (3) use of resistant rootstocks, and (4) balanced nutrient application (Chopra 2022).

Genetic Resistance

Genetic resistance refers to those heritable features of a host plant that suppress or retard the development of a pathogen. Genetic resistance is considered as a major form of biological control of biotic stresses. The main features of genetic resistance are mentioned below:

- Genetic resistance is governed by nuclear genes or cytoplasmic genes or both. Genetic resistance is passed down from one generation to the next through the host organism's genetic material, typically encoded in its DNA.

- Genetic resistance is measured about susceptible varieties or genotypes. Genetic resistance is often pathogen-specific or at least specific to a group of related pathogens. Different genes or genetic mechanisms may confer resistance against distinct pathogens.
- Genetic resistance involves the recognition of pathogen molecules, such as effector proteins, by host receptors (R proteins). This recognition triggers a cascade of molecular events leading to the activation of defense mechanisms.
- Many cases of genetic resistance involve the induction of a hypersensitive response (HR), where infected plant cells undergo programmed cell death at the infection site. This restricts pathogen growth and spreads the immune response.
- Breeding for disease and insect resistance differs from breeding for higher yield. There is a triangular interaction of the *host–pest–environment* in resistance breeding (Fig. 1).

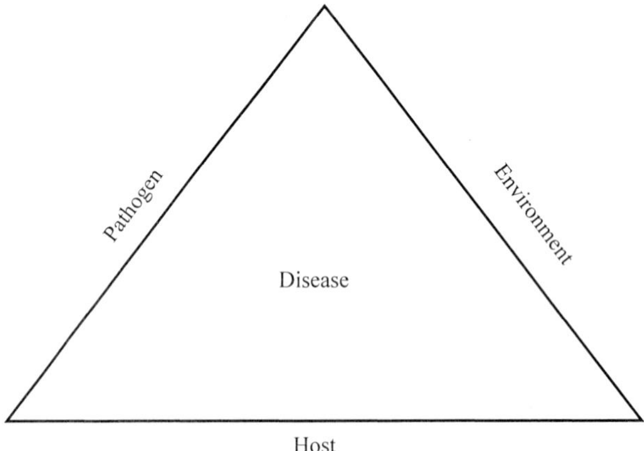

Fig. 1: A disease triangle is used to illustrate the interactions among three essential components that contribute to the development of a disease in plants. These components are the host plant, the pathogen (disease-causing microorganism), and the environment. The disease triangle helps explain how these factors interact to influence the occurrence and severity of a plant disease.

- Genetic resistance is often characterized by a "gene-for-gene" interaction, where specific host resistance genes (R genes) recognize corresponding pathogen avirulence genes (AVR genes). This interaction determines the outcome of the host–pathogen interaction.

Vertical and Horizontal Resistance

Van der Plank (1966) introduced the terms and the concept of vertical and horizontal resistance. A brief description and relevance of each term is presented below:

- *Vertical or specific resistance*: It is also known as *race-specific, pathotype-specific,* or simply *specific resistance*. Vertical resistance is the specific resistance of a host to a particular race of a pathogen. It is generally determined by major genes and is characterized by pathotype-specificity. *Pathotype-specificity* denotes that the host carrying a gene for vertical resistance is attacked by only that pathotype, which is virulent toward that resistance gene: to all other pathotypes, the host will be resistant. This resistance is typically controlled by a single major gene or a small set of genes and therefore, is referred to as oligogenic resistance. It provides effective protection against a specific pathogen race while often being ineffective against other races or strains of the same pathogen. The interaction between the rice plant and the rice blast fungus (*Magnaporthe oryzae*) is a classic example of vertical resistance. Different races of the fungus can overcome specific resistance genes in rice varieties. For instance, the *Pi-ta* gene in rice confers resistance against a specific race of the rice blast fungus that carries the corresponding avirulence gene, AVR-*Pita*. If a new race of the fungus lacking the AVR-*Pita* gene emerges, plants with the *Pita* gene become susceptible (Singh 2018).
- *Horizontal or general resistance*: It is the resistance of a host to all the races of a pathogen, also known as *race-nonspecific, pathotype-nonspecific, minor gene resistance*, and *partial or general resistance.* Horizontal resistance is controlled by polygenes. This resistance does not protect plants from infection, but it does reduce the susceptibility of the host plant and slows down the establishment of the pathogen. It is also known as *field resistance*. Thus, horizontal resistance is incomplete but is permanent as some resistance in the host will always be present to all the races of the pathogen (Table 1).

Table 1: A comparison of the various features of vertical and horizontal disease resistance

Feature	Vertical resistance	Horizontal resistance
Type of resistance	Specific, race-specific	Nonspecific, broad-spectrum
Genetic basis	Controlled by few major genes	Often involves multiple minor genes
Recognition mechanism	Gene-for-gene interaction	No specific recognition required
Specificity	Highly specific to a particular pathogen race	Effective against multiple pathogens
Effectiveness range	Effective against a limited range of pathogen races	Effective against a wide range of pathogens
Durability	Prone to breakdown due to pathogen evolution	More durable, less likely to break down
Deployment strategy	Requires continuous development of new resistant genes	The sustainable, long-term solution
Host plant varieties	May require constant development of new resistant varieties	Fewer new varieties may be needed
Pathogen evolution	Rapidly overcome by new pathogen races	Slower development of resistance-breaking strains
Resistance mechanism	Typically involves R gene and avirulence (AVR) gene interaction	Enhanced defense responses or immunity
Development time	Involve the rapid development of new pathogen races	Generally, takes longer to develop new strains
Examples	Rice blast resistance in rice	General disease resistance in some crops

Gene-for-gene Hypothesis

The concept of gene-for-gene hypothesis, also known as the "Flor hypothesis" was proposed by the Dutch plant pathologist, Jan Hendrik Flor in 1956 based on his studies of host–pathogen interaction in flax or linseed (*Linum usitatissimum)* for rust caused by *Melampsora lini*. The hypothesis explains the interactions between plants and their pathogens.

The hypothesis states that for each gene controlling resistance in the host, there is a corresponding gene controlling pathogenicity in the pathogen.

- The plant gene (R gene): The plant possesses certain resistance genes (R genes) that code for proteins or other molecules capable of recognizing specific pathogen-derived molecules, known as AVR genes or effectors. These AVR genes are produced by the pathogen and play a role in the infection process.
- The pathogen gene (AVR gene): The pathogen, in turn, possesses AVR genes that code for effectors, which are molecules that enable the

pathogen to successfully infect the plant. These effectors interact with or manipulate the plant's cellular processes to facilitate infection.

The interaction between a specific R gene in the plant and its corresponding AVR gene in the pathogen leads to a recognizable recognition and defense mechanism. The genotype of the host and pathogen determine the disease reaction. When genes in the host and pathogen match for all the loci, then only the host will show a susceptible reaction. If some gene loci remain unmatched, the host will show a resistant reaction.

This hypothesis applies to most of the host–parasite systems in which resistance is conditioned by major (vertical resistance) genes and virulence increases in a stepwise manner. It also states that during their evolution, the host and parasite develop a complimentary gene system so that "for each gene-conditioning rust reaction in the host, there is a specific gene-conditioning pathogenicity in the parasite" (Flor 1956) (Table 2).

Table 2: A simple scheme to explain host-pathogen interaction illustrating the gene-for-gene relationship.

Plant gene (R gene)	**Pathogen gene [avirulence (AVR gene]**	**Outcome**
Present	Present	Plant recognizes pathogen; resistance
Present	Absent	No recognition; susceptibility
Absent	Present	No recognition; susceptibility
Absent	Absent	No recognition; susceptibility

Gene-for-gene Relationship

The gene-for-gene relationship between a host and its pathogen was postulated by Flor in 1951 based on his work on linseed (*L. usitatissimum*) rust caused by *M. lini.* Subsequent studies have shown that the gene-for-gene relationship holds in most of the cases studied extensively, and is now widely accepted. It has been found that for every resistance gene present in the host, the pathogen has a gene for virulence. Susceptible reaction would result only when the pathogen can match each of the resistance genes present in the host with the appropriate virulence gene. If one or more resistance genes are not matched by the pathogen with the appropriate virulence gene, the resistant reaction will result. In most pathogens, virulence is recessive to avirulence (Table 3).

Table 3: Gene-for-gene relationship in several host-parasite systems (Day, 1974)

Disease	System
Rust	*Linum–Melampora lini*
	Triticum–Puccinia graminis tritici
	Triticum–Puccinia striiformis
	Triticum–Puccinia recondita
	Zea–Puccinia sorghi
Smut	*Triticum–Ustilago tritici*
	Hordeum–Ustilago hordii
Bacteria	*Gossypium–Xanthomonas malvacearum*
Virus	*Lycopersicon–TMV (tobacco mosaic virus)*
	Lycopersicon–spotted wilt
	Solanum–virus X

Mode of Inheritance of Genetic Resistance

There are three different modes of inheritance, viz., (1) oligogenic, (2) polygenic, and (3) cytoplasmic.

Oligogenic Resistance

When disease resistance is governed by one or a few major genes, and resistance is generally dominant to the susceptible reaction, it is known as oligogenic resistance. It is race specific and provides resistance against one or a few races of pathogen. There are several examples of oligogenic resistance to disease (Table 4). In some cases, oligogenic resistance is altered by modifier genes, e.g., bunt resistance in wheat. Transfer of oligogenic resistance from one host to another is simple.

Table 4: Oligogenic resistance.

Crop plant	Disease	Oligogenic resistance genes	Reference
Rice	Blast	Pi21, Pi9	Li *et al.* 2016; Kumar *et al.* 2018
Tomato	Late blight	Ph-3, Ph-4	Danan *et al.* 2016; Vossen *et al.* 2016
Wheat	Leaf rust	Lr34, Lr67	Singh *et al.* 2015; Saintenac *et al.* 2018
Potato	Potato virus Y	Rysto	Valkonen *et al.* 2008

Polygenic Resistance

When disease resistance is governed by several genes each having a small and generally additive effect. In the case of polygenic resistance, disease reaction is not classifiable into clear-cut resistant and susceptible classes. The polygenes show both additive and nonadditive effects, and there is a large environmental effect, as is the case with other quantitative traits. This type of resistance is generally more durable than oligogenic resistance because it involves several features of the host plant. Transfer of polygenic resistance from one host to another is more difficult than oligogenic resistance. There are several examples of polygenic disease resistance. Some examples are included in Table 5.

Table 5: Polygenic resistance.

Crop plant	**Disease**	**Polygenic resistance genes**	**References**
Rice	Blast (*Magnaporthe oryzae*)	QTLs: Pi21, Pi9, Pi5, Pi2/Piz	Khush *et al.* 1990
Wheat	Leaf rust (*Puccinia triticina*)	QTLs: Lr34/Yr18, Lr46/Yr29	Singh *et al.* 2015
Tomato	Late blight (Phytophthora infestans)	QTLs: Ph-2, Ph-3, Ph-4	van der Vossen *et al.* 2000
Maize	Gray leaf spot (*Cercospora zeae-maydis*)	QTLs: qGLS1, qGLS3	Prasanna *et al.* 2003
Barley	Powdery mildew (*Blumeria graminis f. sp. hordei*)	QTLs: Mla, Ror1	Zhang *et al.* 2014

Cytoplasmic Resistance

Sometimes disease resistance is governed by cytoplasmic gene(s) or plasmagene(s). For example, maize strains having the T male sterile cytoplasm (cms-T) are extremely susceptible to southern leaf blight (*Harpophora maydis*), and resistance to yellow virus in sugarbeets.

Cases of cytoplasmic inheritance of disease resistance are rare.

Systemic Acquired Resistance

Systemic acquired resistance (SAR) is a defense mechanism in plants that involves the activation of a broad-spectrum immune response following an initial encounter with a pathogen or other stress-inducing factors. SAR enables plants to enhance their resistance to a variety of pathogens, even those that are different from the one that triggered the response. This phenomenon is a crucial aspect of the plant's innate immune system and contributes to its ability to fend off infections. SAR relies on signaling molecules like salicylic acid to activate defense genes. Instead of specific resistance, SAR readies the plant to fight multiple pathogens. This process involves priming, where the plant preps its defenses. SAR enhances plant immunity and has practical applications in crop protection. It is a long-lasting systemic resistance that is often effective against viral, bacterial and fungal pathogens, and is induced by such pathogens that cause a necrotic reaction, which may range from HR to necrotic disease lesions. Development of SAR is associated with the activation of SAR genes which seem to be different in dicots and monocots plants.

Systemic acquired response has been observed in the interaction between tobacco plants (*Nicotiana* spp.) and the tobacco mosaic virus (TMV). When a tobacco plant is infected with TMV in one part of its leaves, it triggers a systemic response throughout the entire plant. This systemic response involves the production and transport of signaling molecules like salicylic acid, which activate defense genes in unaffected parts of the plant. As a result of this systemic response, the uninfected parts of the tobacco plant become more resistant to not only TMV but also to other pathogens. This broad-spectrum resistance is a characteristic of SAR (Métraux 2002).

Methods of Breeding for Disease Resistance

The commonly used methods of breeding for disease resistance are described in the following text.

Selection

The easy and quickest method of resistant breeding is a selection of disease-resistant plants from a commercial variety. This method has been used extensively in the past, but at present has limited usefulness for crop improvement. The selection method is a fundamental approach in plant breeding that involves choosing and propagating plants with desirable traits to create new generations of plants with improved characteristics. Examples include resistance to apple scab, a fungal disease in apples, to mildew and Pierce's disease in grapes, and to *Periconia* root rot in sorghum.

Recurrent Selection

Recurrent selection is a breeding strategy aimed at enhancing desired traits, such as disease resistance, in plants over multiple generations. This approach involves systematically selecting and crossing plants that display the desired trait, such as resistance to a specific disease, to accumulate favorable genes in the population. This process is typically carried out over several cycles, with each cycle involving the selection of the best-performing individuals from the previous generation and their subsequent interbreeding. For example, maize varieties with improved resistance to northern leaf blight have been developed by employing recurrent selection.

Introduction

It involves the transfer of plants with desirable disease-resistance traits from one region to another, to broaden the genetic base for breeding efforts to develop new disease-resistant crop varieties. This is a relatively simple and quick means of developing resistant varieties. However, this method offers some limitations which are that the introduced varieties may not necessarily perform well in the new area, may become susceptible to the concerned disease in the new area, and may be susceptible to other races of the concerned disease or other diseases prevailing in the new area. For example, the American cotton varieties introduced in India were susceptible to red blight. Similarly, Kenya wheat introduced in India was rust resistant, but it was highly susceptible to loose smut.

Plant introduction has proven to be a valuable approach for disease control. For example, Kalyan Sona and Sonalika wheat varieties were introduced from CIMMYT, Mexico, and were rust-resistant. Introductions also serve as sources of resistance in breeding programs. For example, African pearl millet introductions have been used for developing Downey mildew-resistant male sterile lines (Tift23 A cytoplasm) for use in hybrid seed production.

Mutation

Both spontaneous and induced mutations serve as valuable sources of disease resistance. The acquisition of resistance against diseases has been accomplished in numerous crops by employing induced mutations. Some examples are resistance to powdery mildew in barley, Victoria blight and crown rusts in oats, flax rust in flax, and leaf and stem rust in groundnut.

Hybridization

Hybridization is the most common method of breeding for disease resistance used when resistant genes are available either in the germplasm or in wild species of crop plants. Transfer of disease resistance from an agronomically undesirable variety to a susceptible but otherwise desirable variety is achieved by the *backcross method* and this method is commonly used when the resistance is governed by oligogenes. Combining disease resistance and some other desirable characters of one variety with the superior characteristics of another variety is achieved by the *pedigree method.*

Pedigree Method

This method is used when the resistance is governed by polygenes and the resistant variety is an adapted one that also contributes some desirable agronomic traits. In this technique, crosses are made among parents (one with the resistance trait and some others with precise agronomic traits), and the individual plant life is decided on for resistance from the F2 generation. This method has been predominantly used in creating a large majority of disease-resistant commercial varieties, e.g., Kalyan Sona, Sonalika, Malviya12,

Malviya37, Malviya234, Malviya 206 and several other wheat varieties, and cotton variety Laxmi resistant to red leaf blight.

Backcross Method

This method proves to be valuable for transferring genes associated with resistance from less favorable varieties in terms of agronomic traits to susceptible varieties that are widely adaptable and highly sought after. Generally, at the culmination of the backcross program, approximately four to six successive backcrosses are conducted. Following this, the offspring undergo self-pollination, and plants exhibiting resistance are selectively chosen. Offspring originating from different homozygous resistant plants with similar agronomic traits are typically pooled together to create a novel variety resistant to disease. This new variety closely resembles the original recurrent parent in all aspects except for its disease resistance. Consequently, extensive yield trials are often not required before its official release.

Somaclonal Variation

Obtaining disease-resistant somaclonal variants can be achieved through the following methods. Firstly, plants that have been regenerated from cultured cells or the offspring of such plants are exposed to disease testing, and plants displaying resistance are singled out (screening). Secondly, cultured cells are carefully chosen for their resistance to toxins or culture filtrates generated by the pathogen, and subsequently, plants are regenerated from these selected cells (cell selection). Generally, these plants also demonstrate resistance to the specific disease in consideration. The cell selection approach is particularly promising when the disease development is influenced by toxins.

Genetic Engineering

Genetic engineering for disease resistance in plants involves the intentional manipulation of an organism's genetic material to introduce genes that confer resistance to specific diseases. This is achieved through the insertion of genes from other organisms, often from naturally resistant species, into the target plant's genome. These genes encode proteins that help the plant recognize and combat pathogens, thus enhancing its ability to fend off diseases. In this approach genes expected to confer resistance are isolated, cloned, and transferred into the crop in question. Genetic engineering offers precision and rapidity in developing disease-resistant plants, as it allows scientists to target specific genes of interest.

Gene pyramiding is a plant breeding strategy where multiple genes conferring resistance to different diseases are intentionally combined into a single plant variety. This approach aims to enhance the plant's overall resistance to a broader spectrum of pathogens by leveraging the combined effects of multiple protective genes. Gene pyramiding increases the durability and effectiveness of disease resistance in crops, reducing the risk of pathogen adaptation and improving long-term crop health and yield, e.g., bacterial leaf blight resistance in rice.

Gene deployment refers to the deliberate introduction and utilization of specific genes, often those conferring desirable traits like disease resistance or improved yield, in plant or animal populations through breeding or genetic modification. This process aims to enhance the targeted population's performance, productivity, or resilience to challenges such as pests, diseases, or environmental stresses.

Advanced Breeding Strategies for Disease Resistance

Conventional breeding faces various challenges, including a time-consuming and labor-intensive selection process, complexities in selecting suitable genotypes due to the quantitative nature of many agronomic traits, and the need for multiple generations of crossing, selfing, and resistance testing. Furthermore, linkage drag is a common setback in conventional breeding, where undesired agronomic traits are unintentionally transferred due to the close linkage with resistance loci. Recent breakthroughs in molecular genetics have introduced innovative avenues for breeders to craft the varieties of tomorrow, known as molecular breeding.

Marker-assisted Selection

Marker-assisted selection (MAS) is a powerful technique employed in plant breeding to enhance disease resistance. This approach involves using molecular markers, which are specific DNA sequences associated with target genes for disease resistance, to streamline the selection process. By identifying plants that possess these markers, breeders can efficiently select individuals with desired disease resistance. In wheat, MAS has been used to incorporate multiple resistance genes against stem rust, resulting in varieties like Thatcher+Lr34 and Thatcher+Lr67. MAS-based early generation selection not only selects suitable gene combinations but also ensures a high probability of retaining superior breeding lines (Eathington *et al.* 1997; Ragimekula *et al.* 2013).

Marker-assisted Backcrossing

It is a refined application of marker-assisted selection (MAS) that focuses specifically on introgressing disease resistance genes into elite plant varieties through repeated backcrossing. This technique helps overcome the challenges of linkage drag and streamline the process of incorporating resistance genes while retaining the desirable agronomic traits of the recurrent parent. In MABC, molecular markers linked to the disease resistance gene are used to track the presence of the resistance allele during the backcrossing process. This allows breeders to efficiently select plants with the target resistance gene while minimizing the inclusion of unwanted genomic regions from the donor parent.

Genomic Selection

Genomic selection for disease resistance in plants is a modern breeding approach that employs genetic information from an individual's entire genome to predict its potential to resist specific diseases. By analyzing numerous genetic markers across the genome and their associations with disease resistance, genomic selection enables breeders to identify plants with enhanced resistance without the need for extensive field testing. This method leverages the cumulative effects of multiple genes contributing to resistance, allowing for a more efficient and accurate selection of plants with improved disease resistance traits. This technique involves analyzing the genetic markers spread across the genome to estimate the genetic potential of plants, enabling breeders to select individuals with superior disease resistance (Poland and Rutkoski 2016).

Clustered regularly interspaced short palindromic repeats (CRISPR)-Cas technology has revolutionized the field of genetics and plant breeding, offering a precise and efficient way to enhance disease resistance in plants. CRISPR-Cas allows scientists to edit specific genes, including those responsible for disease susceptibility, thereby creating plants with improved resistance to various pathogens. It is used to knock out or modify susceptibility genes that pathogens exploit to infect plants. By disabling these genes, plants become less vulnerable to infection. CRISPR-Cas can also be used to introduce specific resistance genes into plants, enhancing their natural defense mechanisms against pathogens. For example, the introduction of a resistance gene from wild tomato species into cultivated tomatoes provides them with enhanced resistance against the bacterial pathogen *Xanthomonas perforans* (Borrelli *et al.* 2018).

Conclusion

Diseases result in substantial economic losses and can lead to human suffering by causing crop failures, and in some cases, even triggering famines. Therefore, resistant plant varieties serve as a critical buffer against potential starvation for humanity. Numerous crops owe their economic viability to resistance against specific diseases, often acquired from closely related species. Resistant varieties offer a cost-effective approach to disease management as they avoid recurring expenses for farmers. This economic advantage makes them a particularly viable option for disease control, especially in developing nations. Additionally, these varieties eliminate the need for fungicides, contributing to reduced environmental contamination, and can be seamlessly integrated with other disease management strategies.

The primary objective of breeding for disease-resistant plants is to sustain productivity by mitigating losses caused by diseases. In contrast, prioritizing higher yields primarily benefits yield in disease-free conditions. When faced with varying disease intensities, breeding for disease resistance offers superior protection to overall yield compared to the gains achieved by breeding solely for higher yields. Developing resistant plant varieties not only mitigates these risks, but also diminishes reliance on environmentally harmful chemical interventions. By harnessing natural genetic diversity and integrating cutting-edge techniques, disease-resistant crops bolster resilience, reduce economic burdens on farmers, and contribute to a more ecologically balanced and secure food production system for present and future generations. Balancing resistance with other essential traits, such as yield and quality, poses a dilemma, potentially leading to trade-offs. Ethical concerns related to genetically modified organisms and regulatory hurdles can slow down the development and deployment of resistant varieties. Moreover, the effectiveness of resistance might vary depending on environmental factors, requiring tailored solutions for different regions. These challenges necessitate a multidisciplinary approach, integrating traditional breeding practices with modern genetic technologies, to create enduring disease-resistant plant varieties.

References

Agrios, G. 2005. Plant Pathology 5th ed. Amsterdam: Elsevier Academic Press. pp. 26-27, 398-401.

Borrelli, V.M., Brambilla V., Rogowsky P., Marocco A., Lanubile, A. “The Enhancement of Plant Disease Resistance Using CRISPR/Cas9 Technology.” Frontiers in Plant Science 9 (2018):1245.

Chopra, V.L. 2022. Plant Breeding: Theory and Practices: 2nd Restructured Edition. New Delhi: New India Publishing Agency.

Choudhary, A., Kumar A., Kaur H., Pandey V., Singh B., Mehta, S. 2022. Breeding Strategies for Developing Disease-resistant Wheat: Present, Past, and Future. Cereal diseases: Nanobiotechnological Approaches for Diagnosis and Management Singapore: Springer Nature Singapore. pp. 137-61.

Day, P.R. 1974. Genetics of Host–Parasite Interaction. San Francisco: WH Freeman and Co.

Deng Y., Ning Y., Yang D.L., Zhai K., Wang G.L., He Z. “Molecular Basis of Disease Resistance and Perspectives on Breeding Strategies for Resistance Improvement in Crops.” Molecular Plant 13, 10 (2020):1402-19.

Mehrotra, R.S., Aggarwal A. 2017. Fundamentals of Plant Pathology, 1st ed. India: McGraw Hill Education.

Métraux, J.P., Nawrath C., Genoud T. (2002). "Systemic Acquired Resistance." Advances in Botanical Research vol. 37. United States: Elsevier. pp. 35-51.

Poland, J., Rutkoski J. "Advances and Challenges in Genomic Selection for Disease Resistance." Annual Review of Phytopathology 54 (2016):79-98.

Ragimekula, N., Varadarajula N., Mallapuram S., Gangimeni G., Reddy R., Kondreddy H. "Marker Assisted Selection in Disease Resistance Breeding." Journal of Plant Breeding and Genetics 1, 2 (2013): 90-109.

Randhawa, M.S., Lan C., Basnet B.R., Bhavani S., Huerta-Espino J., Forrest K.L., et al. "Interactions among Genes Sr2/Yr30, Lr34/Yr18/Sr57, and Lr68 Confer Enhanced Adult Plant Resistance to Rust Diseases in Common Wheat (*Triticum aestivum L.*) Line 'Arula'." Australian Journal of Crop Science 12, 6 (2018):1023-33.

Sharma, S.K., Sharma D., Meena R.P., Yadav M.K., Hosahatti R., Dubey A.K., et al. 2021. Recent Insights in Rice Blast Disease Resistance. Blast Disease of Cereal Crops: Evolution and Adaptation in Context of Climate Change Germany: Springer International Publishing. pp. 89-123.

Singh, B.D. 2018. Plant Breeding Principles and Methods, 11th ed. New Delhi: Kalyani Publishers.

Singh, P. 2018. Essentials of Plant Breeding, 7th ed. New Delhi: Kalyani Publishers.

Valkonen, J.P.T., Wiegmann K., Hämäläinen J.H., Marczewski W., Watanabe K.N. "Evidence for Utility of the Same PCR□based Markers for Selection of Extreme Resistance to Potato Virus Y Controlled by Rysto of Solanum stoloniferum Derived from Different Sources." Annals of Applied Biology 152, 1 (2008):121-30.

Van Der Plank, J.E. "Horizontal (Polygenic) and Vertical (Oligogenic) Resistance against blight." American Potato Journal 43 (1966):43-52.

4

Breeding Strategies for Insect Resistance

Radhika Shekhawat[1], Himanshi Jhalora[2], Parul Gupta[3], Dharm Veer Singh[4] and Kamaluddin[5]

[1–3]*Department of Genetics and Plant Breeding, Maharana Pratap University of Agriculture and Technology, Udaipur, Rajasthan, India*

[4,5]*Department of Genetics and Plant Breeding, Banda University of Agriculture and Technology, Banda, Uttar Pradesh, India*

Abstract

Our natural ecology comprises insects, which are known to have both advantageous and detrimental consequences. Insect pests are the main obstacles in maximizing agricultural crop output, since they directly or indirectly result in significant crop loss. The intricate relationship that phytophagous insects have with their hosts is the result of a protracted and ongoing evolutionary process. Because of this, it is impossible to completely eradicate them, but there is a hopeful fix for this issue, i.e., utilizing insect-resistant crop varieties. Insect-resistant kinds are those that, with the same starting infestation level and similar environmental conditions, yield considerably higher yields of high-quality fruit than susceptible varieties. By boosting yield and reducing pesticide use, these varieties had an enormous economic benefit. There are a number of highly effective insect-resistance mechanisms that the targeted insect pest cannot simply overcome. This chapter discusses various breeding strategies such as introduction, selection, and hybridization as well as some modern breeding approaches that have been used to develop many insect-resistant crop varieties that have significantly improved the world's food supply, by reducing carbon inputs through reduced pesticide use. By expressing bacterial endotoxins and producing Bacillus thuringiensis (Bt) crops, genetic engineering also aids in the production of insect-resistant crops. Furthermore, RNA interference (RNAi) and genome editing by CRISPR/Cas9 (clustered regularly interspaced short palindromic repeats/CRISPR associated 9) offer new avenues for the production of insect-resistant crops. Researchers have also devised effective screening tools for selecting pest-resistant crop types, which was formerly the most difficult step in a breeding process. In this chapter, emphasis has been laid on insect resistance mechanisms, breeding tactics, and screening techniques for selection of insect-resistant crop varieties.

Keywords: *Bacillus thuringiensis, breeding strategies, genetic engineering, insect resistance*

Introduction

Insect attack is a major source of agricultural production loss. Currently, 20%–40% of worldwide crop yield is lost each year owing to insect and pest infestation (USDA 2023). Insect and pest damage to crops ranges from 50% in wheat to >80% in cotton. Cotton is attacked by a dozen primary insect pests, resulting in the application of almost 60% of all insecticides used globally. The majority of insect species have a well-developed system for host selection. Polyphagous pests are insect pests that feed on multiple crop species and have a large host range.

Previously, it was believed that chemical control measures would play a significant role in controlling or perhaps eliminating insect pests. However, this was not the case because it resulted in the emergence of pesticide-resistant biotypes of insects that were no longer affected by pesticides. It also increased production costs, weakened natural adversaries of insect pests, and polluted the environment. The increasing difficulties of pest control have highlighted the vital necessity for creating insect resistance breeding tactics to fight the adverse consequences of chemical control measures.

Insect resistance is defined as the trait of a host crop that makes it less susceptible to insect pests than other varieties of the same crop. If a variety is pest-resistant, it provides higher yields of higher quality than susceptible types. The economic crop loss caused by these insects is determined not only by the pest attack population, but also by the plant's sensitivity to the infestation, stage of plant growth, and duration of the attack. It is also worth noting that abiotic and biotic forces naturally govern these insects.

Breeding crop plants resistant to insects is a complex problem, so more information about the interaction between the host and insect is required to allow for better knowledge for both breeders and entomologists to carry out efficient breeding programs. It is well-known that plant resistance to insects is the more economical and highly effective means of reducing damage to crop plants, and minimizing hazards to humans, animals, and natural predator.

Plant resistance refers to the collective heritable qualities that allow a plant species, race, clone, or individual to diminish the likelihood of an insect pest successfully using that plant as a host. Plant genetic engineering technology allows for the development of insect-resistant crops by the insertion and expression of insect resistance genes. Insect-resistant crop varieties of economic value can be created by using either delta endotoxin coding sequences taken from the *Bacillus thuringiensis* (Bt) bacterium or genes encoding lectins or enzyme inhibitors derived from plant species.

Crop Losses Caused by Insects

Insects are the reason for both quality and quantity losses in agricultural yield. However, a breeder is more concerned about the loss of quantity. They reduce yield by draining the cell sap directly or by devouring plant components (tissue-feeding insects). They also spread several plant diseases, which contribute to yield loss. Insect infestation causes stunted plant growth, damage to leaves, stems, branches, flowers, fruits, and seeds, and plant withering. The damage due to the insects may be classified as direct or indirect.

- *Direct damage*: Plant growth is reduced or stunted as a result of damage to the leaves, stems, branches, lower buds, flowers, fruits, seeds, vegetative buds, early defoliation of leaves, and finally withering of plants (Kavitha and Reddy 2012).
- *Indirect damage*: Many insects, such as aphids, mites, whiteflies, leaf/planthoppers, and others, spread plant viruses, acting as pathogen vectors. In addition, insect damage makes plants more sensitive to fungal and bacterial diseases (Kavitha and Reddy 2012).

Insects are estimated to cause a 14% loss in crop yield on a global scale (Cramer 1967). Temperature and climate also play an important role in insect activity, as losses due to insect pests are much higher in tropical and subtropical climates than in temperate climates because temperate climates have low temperature and humidity, which makes it challenging for most insects to successfully carry out their metabolic processes and reproduction.

The loss of yield can occur both in the field (preharvest) and in storage (postharvest) as a result of field and storage pests, respectively. Estimates of yield loss vary based on cultivar, pest population density, timing of pest attack in relation to crop phenology, and cultural practices followed.

In the instance of cotton, global cotton production was 118.5 million bales in 2018–2019, which is 4.2% lower than the previous year. Insects and diseases have caused 15%–30% economic losses in cotton production and up to 50% losses due to direct damage or disease transmission in plants (Zafar *et al.* 2020). As a result, Bt crops are introduced that have been genetically altered (modified) to contain the endospore (or crystal) Bt toxin, which makes them resistant to specific insect pests.

The economic impact of Bt crops is significant in some countries since the efficacy of insect-resistant crops through Bt has been shown commercially and is an attractive economic alternative to synthetic insecticides. The synthetic insecticide market is worth around $8.11 billion USD each year; 30% of these insecticides are applied to vegetables and fruits, while 23% and 15% are used to protect cotton and rice, respectively. The production of Bt cotton resulted in a 49.8% reduction in insecticide use worldwide, with Mexico and China leading the way with 77% and 65% reductions, respectively, followed by Argentina (47%), India (41%), and South Africa (33%) (Bakhsh *et al.* 2015).

Degree of Resistance

Resistance is a relative quality that can only be examined in relation to susceptibility; it cannot be measured in isolation. Resistant plants are those that, given the same environmental conditions, are naturally less harmed or infested than others (Painter 1958). There may be varying degrees of resistance which can be grouped into five categories: immunity, high resistance, low resistance, susceptibility, and high susceptibility.

1. ***Immunity***: This is a relationship between a plant and an insect in which the insect is unable to harm the plant under any circumstances. A plant that is immune might thus be referred to as a non host plant.
2. ***High resistance***: In this instance because of characteristics of the host plant, the specific insect only moderately damages the surroundings.
3. ***Low resistance***: Low resistance, also known as partial resistance, is the degree of resistance at which the host suffers less harm from an insect than is typical.
4. ***Susceptibility***: A susceptible response is one in which there is more damage or infestation than is typical.
5. ***High susceptibility***: The degree of damage in this case is significantly higher than average for the crop.

Mechanism of Insect Resistance

Insect pests attack the host plant based on the primary and secondary metabolites found in the plant. The odor and taste of the metabolites are important factors in host plant selection. Because the resistant type lacks the necessary taste and odor for the insect, it interferes with the host selection process.

There are four major classifications for mechanisms of resistance used, which are summarized as nonpreference, antibiosis, tolerance, and avoidance. The first three mechanisms were given by Painter (1951) and the fourth one was added subsequently.

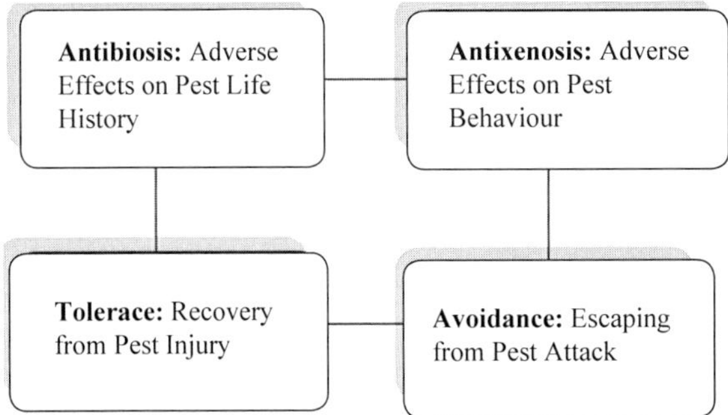

Nonpreference

Nonpreference is one that is not preferred by the insect pest because it is either unattractive or not suited for the insect's oviposition and colonization. It is also referred to as antixenosis (xenos = guest) or nonacceptance. It is caused by the presence of morphological or chemical (allomones/kairomones) elements that affect insect behavior, resulting in poor insect pest establishment on the host crop.

Color, light penetration, hairiness, leaf angle, leaf color, fragrance, and taste are the various plant characters which are all associated with nonpreference (Fehr 1987). For example, the Brown planthopper (BPH) does not prefer rice because of its low asparagine concentration, red pericarp, and purple stigma (Table 1).

Table 1: Nonpreference mechanism for insect resistance.

Host plant character	Imparts resistance to
Trichomes in cotton	Whitefly
Wax bloom on crucifer leaves	Diamond back moth
Hairiness in soybean	Potato leafhopper
Low sinigrin in *Brassica*	Cabbage aphid
Low aspargine in rice	Brown planthopper
Nectariless cotton	Bollworms
Low coumarin in sweet clover	Flying weevil

Nonpreference is not an effective technique of resistance in the case of polyphagous insects since the insect might multiply on the other hosts available and cause damage to the primary crop. For highly polyphagous cotton,

(*Gossypium hirsutum* L.) pests such as *Helicoverpa zea* with a substantial portion of the larval population develops on noncotton alternative hosts. These noncotton hosts may provide *H. zea* with a natural refuge, limiting the evolution of resistance to the Bt-derived Cry1Ac protein seen in Bollguard cotton (Gustafson *et al.* 2006).

Antibiosis

When an insect pest feeds on a resistant plant, the reproduction and growth of that insect pest are slowed due to the negative impacts of host plant. Antibiosis is the process through which the growth of an insect pest is slowed. It may also result in the death of the insect pest. Antibiosis is brought about by physiological, morphological, and biochemical characteristics of the host plant. As we observe in the case of rice cultivars with high silica content, the reproduction of rice stem borer is limited.

Antibiosis affects the insect's life history variables, such as reproduction rate or fecundity and development period, by influencing its physiology. Antibiosis can be caused by the presence of hazardous substances, a lack of vital nutrients, or nutrient imbalance/poor nutrient utilization.

Factors contributing to antibiosis can be classified into two types, *viz.,* chemical and physical factors (Table 2).

Physical factors contributing to antibiosis

- Thick cuticle
- Glandular hairs
- Silica deposits
- Tight leaf sheath

Table 2: Chemical factors contributing to antibiosis.

Chemicals present in plants	**Imparts resistance to**
High gossypol in cotton High tannin in cotton High silica in cotton	Bollworms
High saponin in alfalfa	Spotted aphid and pea aphid
Gummy trichomes exudate in potato	Aphid
High DIMBOA in leaves of maize High aspartic acid and low nitrogen in maize	European corn borer Maize stem borer
Low levels of amino acids and high levels of sugar in peas	Pea aphid
Cucurbitacin in cucurbits	Fruit flies

Tolerance

Tolerance is defined as the host plant's ability to grow and generate higher yields despite pest attack at the same level as the susceptible variety. Tolerant kinds recover faster from insect pests than susceptible varieties because tolerance incorporates various elements such as general vigor, wound healing, and mechanical support in tissues/organs. It is measured in terms of the ability to repair and replace damaged components, the potential for flowering compensation, and healthy leaf growth. Tolerance has little effect on the insect population in terms of selection. However, its main disadvantage is that it is significantly more susceptible to environmental influences than nonpreference and antibiosis.

Avoidance

Avoidance is a mechanism which includes escaping of a variety from insect attack either due to early sowing/early maturity or due to its cultivation in that region where the insect population is very low. For example, the early maturing varieties of cotton escape the pink bollworm infestation which occurs late in the season.

Host Plant Resistance (HPR): Biochemical

Biochemical characters: Biochemical host plant resistance refers to the ability of certain plants to defend themselves against insect pests through the production of chemical compounds that deter feeding, development, or reproduction of the insects. Several secondary metabolites, nutrients and allelochemicals are responsible for the resistance characters against the several insect-pests.

- Nutrient deficient plant acuse antibiotic and antixenotic effects on the insect.
- Antibiosis may result from the absence of certain nutritional substances in the host plants, deficiency of some nutritional materials; imbalance of available nutrients (Table 3).

Table 3: Resistance due to nutritional plant deficiencies

Host-Plant	Nutritional factors	Insect-Pest
Peas	Deficiency of amino acids	Pea aphid, *Acyrthosiphon pisum*
Bean	Lower amounts of carbohydrates and reducing sugars	Mexican bean beetle, *Epilachna varivestis*
Rice	Presence of low quantities of asparagines rice variety	Brown plant hopper, *Nilaparvata lugens*
Barley	Low free amino acids and high surface wax	Cercalaphid, *Rhopalosiphum padi*
Maize	High amount of non- essential amino acid, aspartic acid	Fall armyworm, *Spodoptera frugiperda*

As in maize there are different biochemical defense mechanism that counteract with the various insect pest of maize (Fig. 1).

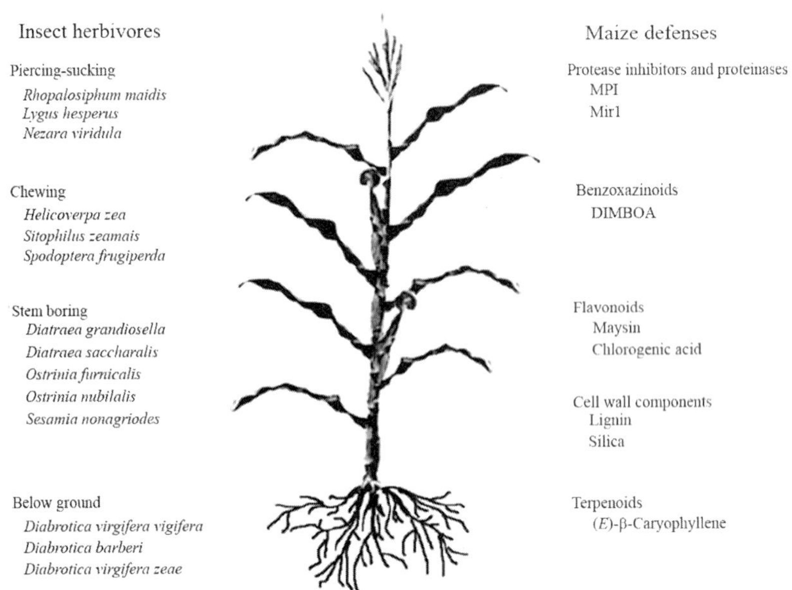

Fig. 1: Maize insect pests and plant defenses that can counteract them.
Source: Meihls *et.al.*, 2012

Breeding Methods for Insect-resistant Varieties

Traditional Methods

- ***Introduction*:** One of the simplest and quickest ways to establish an insect-resistant variety is through introduction. In this case, an introduced variety that is resistant to a certain insect is released for cultivation if it performs well in the new habitat and is agronomically desirable. The transfer of *Daktulosphaira vitifoliae*-resistant grape rootstock from the United States into France, which saved the grape industry from destruction, is the most common example.
- ***Selection*:** Sometimes insect-resistant variants are already present in population, therefore selection for insect resistance may result in an insect-resistant version of that variety. Pureline selection is used for self-pollinated crops. However, because cross-pollinated crops have higher genetic variety than self-pollinated crops, this approach is more effective for them.

Example of selection is resistance to potato leafhopper and spotted alfalfa aphid.

- ***Hybridization***: When the desired insect resistance is present those varieties which are agronomically inferior types, hybridization is a powerful tool for a breeder. To develop a superior insect resistant variety, the resistant strains are crossed with agronomically superior and well-adapted variety which is susceptible to insect pest.
- ***Pedigree method***: When the resistance is governed by either polygenes or oligogenes then pedigree method is suitable to use. If the resistance is governed by oligogenes then the selection is practiced in F2 generation, but when the resistance is governed by polygenes, the selection is practiced in F3 generation.
- ***Backcross method***: In numerous instances, the insect-resistant cultivar is neither superior agronomically nor desirable. In this situation, the backcross approach is applied, with the resistant variety serving as a nonrecurrent parent and an agronomically superior variety serving as a recurrent parent from which the resistance gene is transferred. It is useful when the genes driving resistance are highly heritable as well as when resistance is ruled by only one or a few genes.

Biotechnological Approach

- ***Genetic engineering***: With the invention of superior insect-resistant crop varieties harboring resistance against insect pests, genetic engineering has significantly revolutionized agriculture. The main source of insect-resistant genes has been Bt. The Cry toxin is activated by proteolytic enzymes found in the alkaline juice of the insect's gut (pH 8–10). The active toxin goes through the insect's stomach's peritrophic membrane and then binds to specific receptors on the midgut epithelial cells, creating pores through which the toxin penetrates and swells up the cell until it lyses. The alkaline gut juices then flow into the hemocoel, raising the pH of the hemolymph and ultimately leading to the insect's paralysis and death (Soberón *et al.* 2010).

  For example: Transgenic cotton CH12 expressing *cry2AX1* Genetic engineering of crops for insect resistance showed stable integration and expressed protein constantly for about 90 days after sowing (DAS) with 1.055–1.50 µg/g in fresh leaf with 90% mortality of *Helicoverpa armigera* in T3 plants (Jadhav *et al.* 2020).
- ***RNAi***: RNAi is the process of sequence-specific suppression of gene

expression. RNAi-based insecticidal strategy offers new dimensions for environment-friendly insect pest management in plants (Mamta and Rajam 2017). For example: Silencing of TRPH, ATPD, ATPE, and CHSI genes in soybean resulted in resistant to aphid (Yan *et al.* 2020).

- ***CRISPR-Cas9 (clustered regularly interspaced short palindromic repeats (CRISPR)/CRISPR associated 9)-mediated genome editing for insect control*:** CRISPR/Cas9 is a bacterial defense mechanism against the invading bacteriophages. Cas9 and customizable single guide RNA (sgRNA) are two main important components required for CRISPR-Cas9-mediated genome editing (Talakayala *et al.* 2020). For example, the CRISPR/Cas9 induced mutation in two β-1, 3-glucanase genes of Golden Promise cultivar of barley led to reduction of callose deposition in sieve tubes. Thus, the aphid, *Rhopalosiphum padi*, failed to access the phloem sap and it negatively affected the aphid growth and diminished the host preference in barley (Kim *et al.* 2020).
- ***Marker-assisted selection (MAS)*:** Marker-assisted backcross breeding has been used to successfully incorporate important genes or quantitative trait loci with large effects into commonly established cultivars. It can be used to monitor the presence or absence of insect-resistant genes in breeding populations, and it can be used with traditional breeding methods to generate pest-resistant varieties in the future. Pyramiding various resistance genes with MAS allows breeders to generate broad-spectrum insect resistance. In the case of rice, for example, ASD7 with a BPH resistance gene bph2 was crossed to a susceptible cultivar C418, a japonica restorer line. From a cross between "ASD7" and "C418", 134 F2:3 lines were produced and used to test BPH resistance. In the BC1F1 and BC2F1 populations, both phenotypic and MAS were used (Sun *et al.* 2006).

Screening Techniques for Insect Resistance

A breeder's principal task is to breed for pest resistance. The most important and critical task, however, is to identify such insect-resistant plants during the segregating generation. Field screening, greenhouse screening, laboratory screening, bioassay procedures, and artificial inoculation are the five types of screening approaches (Shashikala 2022).

Field Screening

Test cultivars are grown in fields with natural infestations, either in endemic locations or by using strategies to increase field infestation. Insect attractants, such as fish meal for sorghum shoot fly, can also be applied in the field to attract insects and to boost insect density. The test material can be sowed or planted early, before the adult emerges in a region that was infested previously. In the case of insects that move quickly, such as planthoppers and flies, fiber glass mesh cages can be retained on micro plots and artificially raised insects released at a specific quantity per plant.

Greenhouse Screening

Greenhouse screening gives more reliable results than field tests, as the initial level infestation can be kept more or less uniform for all the test plants. For example, in case of rice, two seedbox screening techniques were compared to determine the levels of resistance of rice cultivars to BPH (Velusamy *et al.* 1986). Seedlings are grown in seed boxes are infested by concerned pest; resistant seedlings are identified and transferred to pots. After a period of recovery, the plants are reinfested by the concerned pest and further selection is made for resistance.

Laboratory Screening

Laboratory screening is the most reliable and a rapid one as it is not affected by any external environmental factors as in the case of field screening and greenhouse screening. Plant tissues are used as insect feeding bioassay for the chewing insects. Plant damage by insects is measured on the basis of area fed, reduced leaf area, loss in dry weight index, and poor chlorophyll concentration.

Bioassay Screening

This technique is used to screen resistance to insect such as *Heliothis* spp. and pink bollworm in cotton. Lyophilized square powder is incorporated in an artificial diet and dispensed into two-ounce plastic containers. Late first or early second instars larvae of *Heliothis* are weighed and placed in the diet cup. By periodic observations, larval survival, larval growth, and percent pupation are recorded. Dates on fecundity and longevity of emerging adults are obtained to screen resistant types (Shashikala 2022).

Artificial Inoculation

Artificial inoculation includes artificial diets that have been developed for rearing of insects that can be made available throughout the year for screening tests. For example in sorghum, the spotted stem borer can be reared on an artificial diet. But if it is not possible to rear insects on an artificial diet, insect colonies can be maintained on natural hosts (shoot fly, head bugs, and midges) under greenhouse conditions (Shashikala 2022).

Conclusion

Plant resistance to insect pests is an important component of integrated pest control because the insects are causing a reduction in agricultural output. Insect resistance does not pose any harm to the environment because no chemical pesticides are used to control insect attack. As a result, breeding for insect-resistant crop species is a cost-effective method of controlling insects in low-value crops. Insect resistance can be paired with host plant characteristics that boost the activities of biological enemies of insect pests, increasing the efficiency of insect resistance and reducing crop loss.

Indeed, the creation of insect-resistant crops represents a significant step forward in agriculture in terms of increasing agricultural productivity and reducing pesticide dependency. As a result, breeders must work harder to accelerate the creation of insect-resistant plants. Plant breeders need to exploit recombination and stack resistant genes to improve the annual genetic gain. However, this is not attainable with standard recombination breeding, since population sizes have to increase exponentially as the number of features picked increases.

Future efforts and resources in insect resistance breeding should concentrate on the development of quick resistance screening procedures, the identification of new sources of resistance, and the incorporation of modern phenomic and genomic technologies into day-to-day breeding.

References

Bakhsh, A., Baloch, F.S., Hatipoglu, R., Ozkan, H. 2015. "Use of genetic engineering: Benefits and health concerns." Handbook of Vegetable Preservation and Processing, 2nd Edition. Edited by Hui Y. H., Evranuz E.Ö. United States: CRC Press. pp. 81-112.

Gustafson, D.I., Head, G.P., Caprio, M.A. "Modeling the impact of alternative hosts on Helicoverpa zea adaptation to Bollgard cotton." Journal of Economic Entomology 99, 6 (2006): 2116-24.

Jadhav, M.S., Rathnasamy, S.A., Natarajan, B., Duraialagaraja, S. Varatharajalu, U. "Study of expression of indigenous Bt cry2AX1 Gene in T3 Progeny of Cotton and its Efficacy Against Helicoverpa armigera (Hubner)." Brazilian Archives of Biology and Technology 63, 2 (2020).

Kavitha, K., Reddy, K.D. "Screening techniques for different insect pests in crop plants." International Journal of Bio-resource and Stress Management 3, 2 (2012): 188-95.

Kim, S.Y., Begtsson, T., Olsson, N., Hot, V., Zhu, L.H., Ahman, I. "Mutations in two aphid-regulated β-1, 3-Glucanase genes by CRISPR/Cas9 do not increase barley resistance to Rhopalosiphum padi L." Frontiers in Plant Science 11 (2020): 1043.

Mamta, B., Rajam, M.V. "RNAi technology: a new platform for crop pest control." Physiology and Molecular Biology of Plants 23, 3 (2017): 487-501.

Painter, R.H. "Resistance of plants to insects." Annual Review of Entomology 3, 1 (1958): 267-90.

Shashikala, M. "Screening Techniques for Insect Resistance." Vigyan Varta 3, 6 (2022): 118-20.

Singh, B.D. 2018. Plant breeding Principles and Methods, 11th edition. New Delhi: Kalyani Publishers. pp. 499-500.

Singh, P. 2018. Essentials of Plant Breeding, 7th edition. New Delhi: Kalyani Publishers.

Soberón, M., Pardo, L., Muñóz-Garay, C., Sánchez, J., Gómez, I., Porta, H., et al. "Pore formation by Cry toxins." Advances in Experimental Medicine and Biology 677 (2010): 127-42.

Sun, L.H., Wang, C.M., Su, C.C., Liu, Y.Q., Zhai, H.Q., Wan, J.M. "Mapping and marker-assisted selection of a brown planthopper resistance gene bph2 in rice (Oryza sativa L.)." Acta Genetica Sinica 33, 8 (2006): 717-23.

Talakayala, A., Katta, S., Garladinne, M. "Genetic engineering of crops for insect resistance: An overview." Journal of Biosciences 45, 1 (2020): 114.

Velusamy, R., Heinrichs, E.A., Medrano, F.G. "Greenhouse techniques to identify field resistance to the brown plant hopper, Nilaparvata lugens(Homoptera: Delphacidae), in rice cultivars." Crop Protection 5, 5 (1986): 328-33.

Yan, S., Qian, J., Cai, C., Ma, Z., Li, J., Yin. M., et al. "Spray method application of transdermal dsRNA delivery system for efficient gene silencing and pest control on soybean aphid (Aphis glycines)." Journal Pest Science 93 (2020): 449-59.

Zafar, M.M., Razzaq, A., Farooq, M.A., Rehman, A., Firdous, H., Shakeel, A., et al. "Insect resistance management in Bacillus thuringiensis cotton by MGPS (multiple genes pyramiding and silencing)." Journal of Cotton Research 3, 1 (2020): 1-13

5

Breeding Strategy for Salinity Stress Tolerance

***Monika S.*[1] *and Shobica Priya R.*[2]**

[1]*Department of Genetics and Plant Breeding, Anbil Dharmalingam Agricultural College and Research Institute, TNAU, Trichy, Tamil Nadu, India*

[2]*Centre for Plant Breeding and Genetics, TNAU, Coimbatore, Tamil Nadu, India*

Abstract

Salinity stress poses a significant threat to agricultural productivity worldwide, impeding crop growth and yield potential. Developing crop varieties with enhanced salinity tolerance is essential for ensuring food security and sustainable agriculture. Salinity stress, resulting from excessive salt accumulation in soil and irrigation water, adversely impacts crop growth, water uptake, and nutrient assimilation. Traditional breeding methods, coupled with recent advancements in molecular and genomic techniques, have enabled the development of crop varieties capable of thriving in saline environments. Genetic diversity serves as the foundation for breeding efforts, allowing the identification and selection of individuals with inherent salinity tolerance traits. Utilization of wild crop relatives and landraces provides a valuable reservoir of genetic variation for salinity stress adaptation. Incorporation of molecular markers into breeding programs expedites the selection of desirable traits. Quantitative trait loci (QTLs) associated with salinity tolerance are identified through techniques such as genome-wide association studies (GWAS) and linkage mapping, facilitating marker-assisted selection. Additionally, recent strides in gene editing technologies like CRISPR-Cas9 (clustered regularly interspaced short palindromic repeats–associated protein 9) offer precise modification of candidate genes responsible for salinity response, accelerating the breeding process. A holistic approach encompassing physiological and biochemical aspects is crucial for accurate phenotypic selection. Traits, such as ion homeostasis, osmotic adjustment, and antioxidative capacity, provide insights into a plant's salinity tolerance potential. The amalgamation of genetic diversity exploration, marker-assisted selection, gene editing, physiological trait evaluation, and interdisciplinary collaboration accelerates the progress toward salinity-tolerant crops.

Keywords: *Salinity stress tolerance, breeding strategy, crop improvement*

Introduction

Abiotic stresses have detrimental effect on plant growth and development, which leads to significant losses in the production of crops. Salinity is a major abiotic stress after drought, which affects the normal growth and development of crops. Generally, salt-affected soils are classified based on the parameters like pH, electrical conductivity (EC), exchangeable sodium percentage (ESP) and sodium absorption ratio (SAR), into saline and sodic soils. Saline soils are those with EC of >4 dSm^{-1} and ESP < 15 and has high concentrations of soluble neutral salts of Ca^{2+}, Mg^{2+}, and Na^{+} chlorides and sulfates (Sharma and Singh 2015). Worldwide, about 1,128 Mha soil is estimated to be affected by salinity and sodicity stresses. Concerned with South Asia, total area of soils affected by salt include 52 Mha out of which, 6.73 Mha lies in India (Sharma *et al.* 2014). With increasing population and decreasing size of the cultivable lands, comes the need of utilizing these kinds of problematic soils for crop cultivation to feed the increasing population. Rather than rectifying these large areas of saline-affected soils using different reclamation strategies, it is efficient to breed varieties which tolerate and grow for saline condition.

In relation to the above context, the plants which have the ability to escape salinity by certain avoidance mechanisms are called as halophytes and those which tolerate them are termed as glycophytes (Afzal *et al.* 2023). Salinity causes excess accumulation of salt in cells which impairs the normal growth and development of crops. It causes osmotic imbalance and oxidative damage, which causes secondary stress and imbalance in ion homeostasis (Horie *et al.* 2011). Saline-tolerant crops have the mechanism, which washes out the excess salts or by maintaining the osmotic potential or by production of antioxidant enzymes which consumes the free radicals and thereby protects the cell organelles from getting harm. Thus, developing crops which tolerate the saline condition with the above-mentioned mechanisms becomes inevitable, which is made possible through different breeding methods and strategies.

Breeding for salinity tolerance primarily involves knowing the effect of salinity stress on normal growth and development of the plant, mechanism which the plant employs to get rid of the excess salts accumulating in the plant cells, genes responsible for the salinity tolerance, breeding methods to be followed for incorporating that gene, and testing the adaptability of identified tolerant genotypes across different areas affected by salinity. There is an ample number of breeding strategies which can be employed to obtain saline-tolerant genotypes. This chapter deals in detail about the mechanism for salinity tolerance and distinct aspects of breeding for salinity tolerance for developing saline-tolerant cultivars.

Effect of Salinity on Crop Growth and Development

Plant salt tolerance involves intricate biochemical and molecular pathways. Understanding physiological, biochemical, and cellular level salt stress aids in gauging salt tolerance and underlying mechanisms. Salinity, the world's oldest and widely distributed challenge, poses a major threat for global crop production. It results in detrimental effects on crop growth by causing nutritional imbalance and lowering the soil osmotic potential. The primary hyperosmotic stress leads to secondary stress, i.e., oxidative damage (Zhu 2001). An increase in soil salinity inhibits the seed germination significantly owing to the combined ion toxicity and high osmotic potential. Various reports stated that a reduction in plant growth has been observed due to salinity stress in tomato, cotton, and sugar beet (Safdar *et al.* 2019). But the tolerance level varies among the cultivars. In sugar beet, leaf number was the least affected trait with a reduction in fresh and dry mass of leaves and roots at 200 mM NaCl (Ghoulam *et al.* 2002).

Effect on Photosynthesis

The salinity effect on photosynthesis depends on crop species and salt concentration. Numerous reports showed a net reduction in photosynthetic capacity with the occurrence of salinity. But a positive association was noticed between salinity stress and photosynthetic rate in *Gossypium hirsutum* and *Asparagus officinalis* (Pettigrew and Meredith, 1994; Faville *et al.* 1999). In contrast, very little association between photosynthetic rate and crop growth has been documented in *Triticum aestivum* and *Triticum repens* (Rogers and Noble 1992). A reduction in photosynthetic rate due to salinity is a result of varied factors:

- Dehydrated cell membranes lower CO_2 permeability. High soil and water salt content leads to elevated osmotic potential, limiting water access. Reduced water potential triggers osmotic stress, halting photosynthetic electron transport through intercellular space shrinkage.
- Salt toxicity, mainly from Na^+ and Cl^- ions, hampers plant growth. Cl^- negatively affects photosynthesis via NO_3^- N uptake inhibition in roots. Reduced NO_3^- N uptake and osmotic stress likely contribute to salinity's photosynthesis-inhibiting impact.
- CO_2 supply drops due to stomatal closure, limiting carboxylation. Stomatal closure conserves water but affects chloroplast function, altering activity. Higher conductance boosts CO_2 diffusion, elevating photosynthesis and crop yields. Stomatal conductance and photosynthetic rate link are not universal (Ashraf 2001). Iyengar and Reddy (1996)

found non-stomatal inhibition due to CO_2 diffusion resistance from mesophyll to chloroplast, plus reduced ribulose-1,5-bisphosphate (RUBP) carboxylase efficiency (Iyengar and Reddy, 1996).

- Increased senescence caused by salt stress
- Shift in enzyme activity due to cytoplasmic structural changes
- Negative feedback by decreased sink activity

Effect on Plant Water Uptake

Plants uptake water and minerals in normal circumstances, due to their increased water pressure than the soil. Under salt stress conditions, the plants cannot uptake water due to the increased osmotic pressure in soil (Kader and Lindberg 2010). Furthermore, lower turgor prompts stomatal closure to conserve water, potentially reducing carbon fixation. Stomatal closure might trigger reactive oxygen species (ROS) generation, superoxide and singlet oxygen, impairing cells via lipid, protein, and nucleic acid damage.

Effect on Ion Homeostasis

Ionic toxicity emerges when an imbalance in salt concentrations occurs and disrupt cellular metabolism. Sodium hinders potassium uptake and enzyme activities, although potassium is vital for maintaining cell turgor, membrane potential, and enzyme function. Sodium concentrations above 10 mM impose stresses on cells. Sodium, akin to potassium, crosses membranes and inhibits enzymes, affecting metabolism. Calcium cations offer protection via signaling pathways that regulate Na^+/K^+ transporters. When sensing salt stress via proteins or enzymes, plants elevate cytosolic calcium, a key secondary messenger for biochemical pathways, aiding salt stress responses. Salinity-induced osmotic and ionic stress can halt crop growth, redirecting energy to water conservation and ion balance. To restore function and photosynthesis, plants must self-detoxify-reducing damage, reestablishing homeostasis, and resuming growth (Zhu, 2001).

Tolerance Mechanisms Against Salinity

Mechanisms which are involved in the salinity tolerance are described in the following text.

Ion Compartmentalization and Homeostasis: Ion Uptake and Transport

Plants, either be halophytes or glycophytes could not tolerate salt concentrations in cytoplasm. The excess salts are either sequestered in vacuole or any older tissues to protect the cytoplasm and organelles from salt injury. Excess sodium is transported through Na^+/K^+ antiporter from vacuole. The genes responsible are salt overly sensitive (SOS) which induces signaling pathway involving SOS1, SOS2, and SOS3 proteins which encodes plasma membrane Na^+/H^+ antiporter, which regulates the cellular Na^+ efflux and its transport from root to shoot. Plasma membrane transporters of histidine kinase family also play an essential role in regulating Na^+/K^+ transportation

Biosynthesis of Osmoprotectants and Compatible Solutes

Compatible solutes are those produced by plants to maintain the osmotic balance within the cells. They are polar, uncharged, and soluble in nature and do not disturb the cellular metabolism. Some of the notable osmoprotectants are listed in Table 1.

Table 1: List of some important osmolytes which alleviates salt stress.

Proline	Glycine betaine	Sugars	Polyols
Cysteine arginine and methionine decreases and proline increases	Involved in osmotic adjustment, stabilizes protein, protects photosynthetic apparatus	• Glucose • Fructose • Trehalose • Starch	• Cyclic (pinitol) • Acyclic (mannitol)

Antioxidative Enzymes

Under stress condition, ROS are produced as a result of disturbed electron transport chain (ETC) in chloroplast and mitochondria, where the oxygen acts as electron acceptor. The ROS produced include $^1O_2$ (singlet oxygen), OH^- (hydroxide radical), O_2^- (superoxide radical) and H_2O_2 (hydrogen peroxide). These ROS damage the cell organelles and the genetic material by causing lipid peroxidation, protein oxidation, and DNA damage and can lead to cellular death. Thus, to combat the oxidative stress, the enzymatic and nonenzymatic antioxidants are produced which prey on the ROS and saves the normal cell structure and function. Flowchart 1 shows the common antioxidants produced in plants during stress condition.

Flowchart 1: Classification of antioxidants in plants.

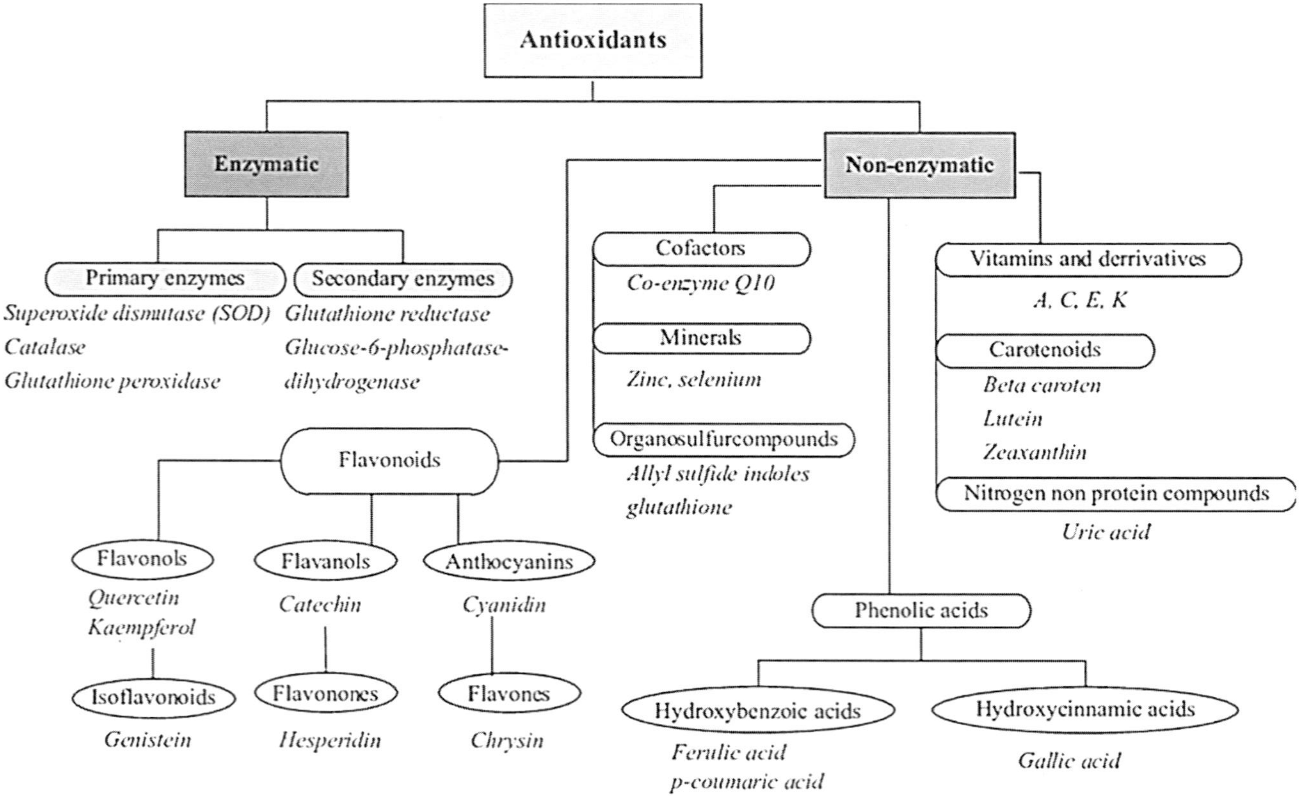

Source: Bunaciu *et al.* 2015.

Production of Polyamines

These are low-molecular-weight compounds which help the plants in normal growth and development starting from cell proliferation, differentiation to seed germination. Commonly found polyamines include the putrescine, spermidine, and spermine. These molecules help in maintaining the integrity of the cell membrane, osmolytes, and induce antioxidants production (Afzal *et al.* 2009).

Hormones Modification

Abscisic acid (ABA), also known as stress hormone, is produced as result of the water deficit which arises due to the salinity around the root (Cabot *et al.* 2009). ABA is noted to partially contribute to the accumulation of K^+ and Ca^+ and the solutes, proline, and other osmolytes to fight against the accumulation of Na^+ and Cl^- (Gurmani *et al.* 2011). It also acts as signal molecule for the activation of several stress responsive genes (Fukuda *et al.* 2006).

Other than ABA, salicylic acid also contributes to tolerate saline stress by preventing the K^+ loss and helps in maintaining the membrane potential. It also accumulates the soluble sugar osmolytes and K^+ ions and found to increase the concentration of photosynthetic pigments to combat the damage caused by salinity (El-Tayeb *et al.* 2005). Brassinosteroids, another class of hormones, also enhance the level of the antioxidant enzymes and helps in alleviating the ROS stress during the saline condition (El-Mashad *et al.* 2012).

Nitric Oxide Generation

Nitric oxide is a major gaseous molecule which is involved in various growth and development processes. During salinity stress, this gaseous molecule wakes up the antioxidant enzymes (Zhang *et al.* 2007) such as catalase (CAT), superoxide dismutase (SOD), ascorbate peroxidase (APX), glutathione reductase (GR) and glutathione peroxidase (GPX) (Mishra *et al.* 2011) which scavenge the ROS and suppresses the MDA thereby preventing lipid peroxidation (Nalousi *et al.* 2012).

Breeding Strategies for Salinity Tolerance

Conventional Breeding Methods

Conventional breeding methods have been successful in developing variations necessary for selection and development of a salt tolerant variety. The preliminary method employed in identifying saline tolerant genotype is screening the germplasm available under saline condition and selecting the tolerant ones. Several studies have been carried out in different crops using germplasm and the saline tolerant genotypes were identified. The identified genotypes are used either as parent as donor for salinity tolerance or used to release as improved cultivar through selection strategies. A lot of crop cultivars and lines resistant to salt have been created by Central Soil Salinity Research Institute in Karnal, e.g., salt-tolerant CSR10, CSR13, and CSR27 rice varieties. However, breeding is limited because of low levels of variation in most crop's gene pools.

Mutation breeding is a breeding method which is employed to create new variation in crop improvement. It employs the use of mutagens like gamma rays, X-rays, ethyl methyl sulfonate (EMS), and sodium azide, which creates changes in the chromosome structure or the nitrogenous bases, which in turn changes the phenotype of the individual. For example, in barley, mutant M4-73-30 was salt-tolerant, different from its susceptible wild cultivar, due to mutation that occurred in the gene responsible for the expression of SOS proteins, thereby making the mutant tolerant (Yousefirad *et al.* 2018). In wheat, using EMS mutation in BARI Gom-25 variety, 76 saline tolerant mutant lines were identified (Lethin *et al.* 2020).

Modern Tools for Developing Salt-tolerant Cultivars

Conventional breeding methods have been successful in developing saline tolerant cultivar, but the rate of success seemed to be low (Afzal *et al.* 2023). The phenomenon of salinity tolerance is complex and has several physiological, morphological, genetic and biochemical responses involved. Current advances in molecular biology help in understanding these mechanisms and it helps in developing saline tolerant cultivars with high precision.

Transgenic approaches can be employed to develop saline-tolerant cultivar. The gene responsible for salinity tolerance can be transferred to the host explant either through *Agrobacterium*-mediated or direct gene transfer.

Markers linked to the gene/quantitative trait loci (QTLs) conferring salinity tolerance are found using different mapping populations. The genes/QTLs for salinity tolerance could be identified using QTL mapping by linkage mapping or association mapping.

Marker-assisted breeding (MAB) has emerged as a powerful strategy for developing saline-tolerant cultivars, offering a targeted and efficient approach to crop improvement. With the increasing challenges posed by soil salinity due to climate change, the need for crops that can thrive under such conditions has become paramount. MAB involves the integration of molecular markers, which are associated with specific traits of interest, into conventional breeding programs. This approach enables breeders to select plants with desired traits more accurately and efficiently, expediting the development of saline-tolerant cultivars. One particular application of MAB is marker-assisted backcrossing

(MABC), which combines the precision of molecular markers with the effectiveness of traditional backcrossing. MABC allows for the rapid transfer of beneficial alleles associated with salt tolerance from a donor parent to an elite recipient cultivar while simultaneously minimizing the introgression of unwanted donor genome segments. Recent advancements in MAB and MABC have demonstrated their potential in enhancing salt tolerance in various crops. For instance, in wheat, MABC has been employed to introgress a major salt tolerance QTL from a wild relative, resulting in improved yield and performance under saline conditions (Platten *et al.* 2020). Similarly, MAB has been utilized in tomato breeding to enhance salt tolerance by selecting for specific molecular markers linked to key salt tolerance genes, thereby improving fruit yield and quality (Munns *et al.* 2020). These examples underscore the efficiency of MAB and MABC in accelerating the development of saline-tolerant cultivars. Moreover, the advent of high-throughput genotyping technologies has revolutionized MAB and MABC approaches. The QTLs for salinity tolerance at seedling and vegetative and reproductive stages in different crops are listed in the review article by Afzal *et al.* (2023). These gene linked markers can be used for MAB to develop saline tolerant cultivar. One of the prominent achievements is the identification of Saltol QTL in rice from Pokkali, which was identified using conventional breeding approach by Thomson *et al.* (2010). This QTL was further used to improve salt tolerance in Bangladesh cultivar, BR 11 using MABC (Gregorio *et al.* 2013).

Additionally, the integration of genomic selection, which involves predicting the breeding value of individuals based on their genotypic data, has further expedited the breeding process by enabling the simultaneous evaluation of numerous genomic regions associated with salt tolerance (Acquaah 2020). The availability of high-quality and informative molecular markers linked to salt tolerance genes is crucial. The choice of markers and their proximity to the target gene influence the success of allele introgression. Furthermore, accurate phenotyping under saline conditions is essential for selecting plants with genuine salt tolerance. Phenotyping methods must accurately reflect the plant's response to salt stress, considering both vegetative and reproductive stages. Integrating genotypic and phenotypic data through sophisticated statistical methods enhances the accuracy of marker–trait associations and facilitates the selection of promising breeding lines (Varshney *et al.* 2017).

Association mapping leverages the natural genetic diversity within a population to identify genomic regions associated with specific traits, such as salt tolerance. Genome-wide association studies (GWAS) takes association mapping to a larger scale by simultaneously assessing thousands to millions of genetic markers across the entire genome. These techniques offer a data-driven approach to dissect the complex genetic architecture of salt tolerance, enabling the identification of candidate genes and markers that can be harnessed. Recent advances in association mapping and GWAS have shown promising outcomes in enhancing saline tolerance in various crop species. In barley, GWAS identified multiple loci associated with salt tolerance-related traits, offering insights into the underlying genetic basis of this complex trait (Mansour *et al.* 2021). Similarly, a comprehensive GWAS in rice revealed novel genomic regions linked to salt tolerance, providing potential targets for further investigation and manipulation (Sandhu *et al.* 2020). Furthermore, the integration of GWAS results with other "omics" data, such as transcriptomics and metabolomics, enhances our understanding of the molecular mechanisms underpinning salt tolerance. The identified candidate genes and markers must undergo rigorous validation and functional characterization to confirm their roles in salt tolerance. In conclusion, association mapping and GWAS have emerged as pivotal approaches in the pursuit of saline-tolerant cultivars. These data-rich techniques enable the dissection of complex traits like salt tolerance by identifying genomic regions associated with the trait of interest. Recent studies in various crop species have highlighted the potential of these methods in uncovering novel genes and markers linked to salt tolerance. Integrating GWAS results with multiomics data provides a comprehensive understanding of the molecular mechanisms governing the plant's response to salt stress. GWAS study by Kumar *et al.* (2015) divulged the molecular mechanism for salinity tolerance.

Genome editing techniques, such as zinc finger nucleases (ZFNs), transcription activator-like effector nucleases (TALENs), and the CRISPR (clustered regularly interspaced short palindromic repeats)/Cas system, have revolutionized agricultural biotechnology by enabling precise modifications in plant genomes. The development of saline-tolerant cultivars, crucial in the face of escalating soil salinity due to climate change, has greatly benefited from these technologies. By precisely targeting specific genes responsible for salt sensitivity and introducing beneficial mutations, these techniques offer the potential to create plants that thrive in high-saline environments without compromising yield or quality. Recent studies have demonstrated the successful application of CRISPR/Cas9 in enhancing salt tolerance in crops like rice (Li *et al.* 2021; Zhang *et al.* 2022). Additionally, TALENs have been utilized to engineer salt tolerance in wheat (Mao *et al.* 2020). As these genome editing methods continue to evolve, the prospect of generating saline-tolerant cultivars with improved agricultural sustainability becomes increasingly promising.

(CRISPR/Cas) system, to develop saline-tolerant cultivars. These techniques offer precise and targeted modifications to plant genomes, thereby unlocking the potential to enhance salt tolerance in crops without compromising yield or nutritional quality. Developing salt-tolerant cultivars through genome editing involves targeting genes associated with salt sensitivity and stress responses. Key genetic components implicated in salt tolerance include ion transporters, hormone signaling pathways, and transcription factors. For instance, the SOS pathway, responsible for maintaining ion homeostasis under saline conditions, has been a primary target for genome editing efforts. Recent research has showcased the remarkable potential of genome editing technologies in enhancing salt tolerance. In rice, the CRISPR/Cas9 system has been employed to engineer mutations in the *OsSOS1* gene, resulting in improved salt tolerance without compromising yield (Li *et al.* 2021). Similarly, the TALENs approach has been used to enhance salt tolerance in wheat by targeting the *TaHKT2;1* gene (Mao *et al.* 2020). These studies highlight the specificity and precision of genome editing techniques, paving the way for the development of saline-tolerant cultivars that can withstand harsh environmental conditions.

Conclusion

Breeding strategies for salinity tolerance have emerged as a crucial avenue in addressing the escalating challenges posed by soil and water salinity in agricultural systems. Through the implementation of conventional and modern breeding techniques, significant progress has been achieved in enhancing the salinity tolerance of various crops, thus improving agricultural productivity in saline-affected regions. The integration of molecular and genetic approaches has facilitated the identification of key genes and mechanisms involved in salinity response, enabling more precise and efficient breeding methods.

Looking ahead, the future prospects of breeding strategies for salinity tolerance are promising. Continued research into the underlying genetic and physiological mechanisms of salt tolerance will enable the development of tailored approaches for different crops. Integrated breeding programs that consider multiple abiotic and biotic stresses, alongside salinity, will be crucial for creating resilient and adaptable crop varieties. Climate change and increasing soil salinization emphasize the urgency of these efforts. Global collaboration and knowledge sharing will play a vital role in enhancing the efficiency of breeding strategies. Thus, breeding strategies for salinity tolerance hold great potential in mitigating the detrimental impacts of salinity on agricultural systems. By harnessing the power of genetics, genomics, and biotechnology, researchers and practitioners can pave the way for a more food-secure and resilient future, where crops can thrive in saline environments and contribute to global food production.

References

Acquaah, G., 2020. Principles of Plant Genetics and Breeding. New York: John Wiley & Sons.

Afzal, I., Munir F., Ayub C.M., Basra S.M.A., Hameed A., Nawaz, A. Changes in Antioxidant Enzymes, Germination Capacity and Vigour of Tomato Seeds in Response of Priming with Polyamines." Seed Science and Technology 37, 3 (2009):765-70.

Afzal, M., Hindawi S.E.S., Alghamdi S.S., Migdadi H.H., Khan M.A., Hasnain M.U., et al. "Potential Breeding Strategies for Improving Salt Tolerance in Crop Plants." Journal of Plant Growth Regulation 42, 6 (2023):3365-87.

Ashraf, M. "Relationships between Growth and Gas Exchange Characteristics in Some Salt-tolerant Amphidiploid Brassica Species in Relation to their Diploid Parents." Environmental and Experimental Botany 45, 2 (2001):155-63.

Bunaciu, A.A., Danet A.F., Fleschin Ş., Aboul-Enein H.Y., "Recent Applications for in vitro Antioxidant Activity Assay." Critical Reviews in Analytical Chemistry 46, 5 (2015):389-99.

Cabot, C., Sibole J.V., Barceló J., Poschenrieder, C. "Abscisic Acid Decreases Leaf Na+ Exclusion in Salt-treated Phaseolus vulgaris L." Journal of Plant Growth Regulation 28 (2009):187-92.

El-Mashad A.A.A., Mohamed H.I. "Brassinolide Alleviates Salt Stress and Increases Antioxidant Activity of Cowpea Plants (Vigna sinensis)." Protoplasma 249 (2012):625-35.

El-Tayeb M.A. "Response of Barley Grains to the Interactive Effect of Salinity and Salicylic Acid." Plant Growth Regulation 45 (2005):215-24.

Faville, M.J., Silvester W.B., Green T.A., Jermyn W.A. "Photosynthetic Characteristics of Three Asparagus Cultivars Differing in Yield." Crop Science 39, 4 (1999):1070-7.

Fukuda, A., Tanaka Y. "Effects of ABA, Auxin, and Gibberellin on the Expression of Genes for Vacuolar H+-Inorganic Pyrophosphatase, H^+-ATPase Subunit A, and Na^{+}/H^+ Antiporter in Barley." Plant Physiology and Biochemistry 44, 5-6 (2006):351-8.

Ghoulam, C., Foursy A., Fares K. "Effects of Salt Stress on Growth, Inorganic Ions and Proline Accumulation in Relation to Osmotic Adjustment in Five Sugar Beet Cultivars." Environmental and Experimental Botany 47, 1 (2002):39-50.

Gregorio, G.B., Islam M.R., Vergara G.V., Thirumeni S. "Recent Advances in Rice Science to Design Salinity and Other Abiotic Stress Tolerant Rice Varieties." Sabrao J Breed Genet 45, 1 (2013):31-41.

Gurmani, A.R., Bano A., Khan S.U., Din J., Zhang J.L. 2011. "Alleviation of Salt Stress by Seed Treatment with Abscisic Acid (ABA), 6-benzylaminopurine (BA) and Chlormequat Chloride (CCC) Optimizes Ion and Organic Matter Accumulation and Increases Yield of Rice ('Oryza sativa'L.)." Australian Journal of Crop Science 5, 10 (2011):1278-85.

Horie, T., Kaneko T., Sugimoto G., Sasano S., Panda S.K., Shibasaka M., et al. "Mechanisms of Water Transport Mediated by PIP Aquaporins and their Regulation via phosphorylation Events under Salinity Stress in Barley Roots." Plant and Cell Physiology 52, 4 (2011):663-75.

Iyengar, E.R.R., Reddy M.P. 1996. Photosynthesis in Highly Salt Tolerant Plants. Handbook of Photosynthesis. United States: Marshal Dekar, Baten Rose. pp. 909.

Kader, M.A., Lindberg S. "Cytosolic Calcium and pH Signaling in Plants Under Salinity Stress." Plant signaling and Behavior 5, 3 (2010):233-8.

Kumar, V., Singh A., Mithra S.A., Krishnamurthy S.L., Parida S.K., Jain S., et al. "Genome-wide Association Mapping of Salinity Tolerance in Rice (Oryza sativa)." DNA Research 22, 2 (2015):133-45.

Lethin, J., Shakil S.S., Hassan S., Sirijovski N., Töpel M., Olsson O., et al. "Development and Characterization of an EMS-mutagenized Wheat Population and Identification of Salt-tolerant Wheat Lines." BMC Plant Biology 20 (2020):1-15.

Li, Y., Wang Y., Zhang H., Zhang Q., Zhai H., Liu Q., et al. "CRISPR/Cas9-mediated Salt Receptor Gene-targeted Mutagenesis Confers Salt Tolerance in Cotton." Plant Biotechnology Journal 19, 6 (2021):1245-7.

Mansour, E., Casas A.M., Gracia M.P., Molina-Cano J.L., Moralejo M., Ciudad F. "GWAS Unveils Key Genes Underlying the Natural Variation of Salinity Tolerance at the Barley Germination Stage." Genes 12, 8 (2021): 1122

Mao, H., Wang H., Liu S., Li Z., Yang X., Yan J., et al. "A Transgene-free Genome Editing System Generates Meiotically Stable Alleles for Crop Improvement." Nature Plants 6, 4 (2020):440-2.

Mishra, S., Jha, A.B., Dubey R.S. "Arsenite Treatment Induces Oxidative Stress, Upregulates Antioxidant System, and Causes Phytochelatin Synthesis in Seedlings." Protoplasma 248 (2011):565-77.

Munns, R., Day D.A., Fricke W., Watt M., Arsova B., Barkla B.J., et al."Energy Costs of Salt Tolerance in Crop Plants." New Phytologist 225, 3 (2020):1072-90.

Nalousi, A.M., Ahmadiyan S., Hatamzadeh A., Ghasemnezhad M. "Protective Role of Exogenous Nitric Oxide against Oxidative Stress Induced by Salt Stress in Bell-pepper (Capsicum annum L.)." American-Eurasian Journal of Agricultural and Environmental Science 12, 8 (2012):1085-90.

Pettigrew, W.T., Meredith W.R. Jr. "Leaf Gas Exchange Parameters vary Among Cotton Genotypes." Crop Science 34, 3 (1994):700-5.

Platten, J. D., Cotsaftis O. 2020. Salinity Tolerance. Wheat Biotechnology. Cham: Springer. pp. 373-94.

Rogers, M.E., Noble, C.L. "Variation in Growth and Ion Accumulation between Two Selected Populations of Trifolium repens L. Differing in Salt Tolerance." Plant and Soil 146 (1992):131-6.

Safdar, H., Amin A., Shafiq Y., Ali A., Yasin R., Shoukat A., et al. "A Review: Impact of Salinity on Plant Growth." Nature and Science 17, 1 (2019):34-40.

Sandhu, N., Subedi P., Macoy D.M., Zheng T., Al-Khatib K. "Genome-wide Association Study of Salt Tolerance at the Reproductive Stage in a Panel of Rice Accessions." Rice 13, 1 (2020):1-13.

Sharma D.K., Singh A. 2015. Salinity Research in India- Achievements, Challenges and Future Prospects. Water Resources Section. Karnal: Central Soil Salinity Research Institute.

Sharma, D.K. Chaudhari S.K., Singh A. 2014. CSSRI Vision 2050. Karnal: Central Soil Salinity Research Institute.

Thomson, M.J., de Ocampo M., Egdane J., Rahman M.A., Sajise A.G., Adorada D.L., et al. "Characterizing the Saltol Quantitative Trait Locus for Salinity Tolerance in Rice." Rice 3, 2 (2010):148-60.

Varshney, R.K., Saxena R.K., Upadhyaya H.D., Khan A.W., Yu Y., Kim C., et al. "Whole-genome Resequencing of 292 Pigeonpea Accessions Identifies Genomic Regions Associated with Domestication and Agronomic Traits." Nature Genetics 49, 7 (2017):1082-8.

Yousefirad, S., Soltanloo H., Ramezanpour S.S., Zaynalinezhad K., Shariati V. "Salt Oversensitivity Derived from Mutation Breeding Improves Salinity Tolerance in Barley via Ion Homeostasis." Biologia Plantarum 62 (2018):775-85.

Zhang, F., Wang Y., Yang Y., Wu H.A.O., Wang D.I., Liu J. "Involvement of Hydrogen Peroxide and Nitric Oxide in Salt Resistance in the Calluses from Populus euphratica." Plant, Cell & Environment 30, 7 (2007):775-85.

Zhang, D., Zhang H., Li T., Chen K., Qiu J.L. "Genome-editing-enabled Salt Tolerance in Rice: Progress and Perspectives." Plant, Cell & Environment 45, 3 (2022):499-510.

Zhao, L., Zhang F., Guo J., Yang Y., Li B., Zhang, L. "Nitric Oxide Functions as a Signal in Salt Resistance in the Calluses from two Ecotypes of Reed." Plant Physiology 134, 2 (2004):849-57.

Zhu, J.K. "Plant Salt Tolerance." Trends in Plant Science 6, 2 (2001):66-71.

6

Plant Tissue Culture: Application in Crop Improvement

Dharm Veer Singh[1], Kamaluddin[2] and Radhika Shekhawat[3]

[1]*Ph.D. Research Scholar, Department of Genetics and Plant Breeding, Banda University of Agriculture and Technology, Banda, Uttar Pradesh, India*

[2]*Professor, Department of Genetics and Plant Breeding, Banda University of Agriculture and Technology, Banda, Uttar Pradesh, India*

[3]*M.Sc. Research Scholar, Department of Genetics and Plant Breeding, MPUAT Udaipur, Rajasthan, India*

Abstract

A collection of in vitro techniques, approaches, and strategies collectively referred to as plant tissue culture make up the field of technology known as plant biotechnology. Tissue culture has been used to enhance the amount of suitable germplasms that are available to plant breeders, improve the health of the planted material, and provide genetic variability from which agricultural plants can be improved. Most crop species have tissue culture techniques available, although many crops, particularly grains and woody plants, still need further improvement. To successfully incorporate particular traits via gene transfer, tissue culture techniques have been applied in association with molecular approaches.

Introduction

One of the biggest problems emerging for the survival of the people in the future is food security. To feed an anticipated nine billion people by 2050, the globe would need to produce roughly twice as much food as it does today. In order to properly accomplish this aim, food production must be elevated sustainably on currently available arable land while addressing the problems brought on by climate change. Over the past five decades, crop yields have steadily increased due to crop breeding programs and better management practices. The rate of yield improvement has, however, reached a plateau (Grassini *et al.* 2013). The noncompounding yearly yield growth of 0.9% for wheat at the present time, world output rises by 38%, is quite small, and is unable to fulfill expected demand levels by 2050. Therefore, the most beneficial course of action is to investigate more effective, higher-yielding crops (Ray *et al.* 2013).

Historically, the genetic improvement of wheat has been accomplished by sexual hybridization between related species, giving rise to a large number of cultivars with excellent agronomic performance and high yields. The main methodology for improving cereal crops is still traditional plant breeding, sometimes in combination with traditional cytogenetic methods (Salina *et al.* 2015). Cereal crops immediately became the top targets for genetic modification due to the worldwide importance of cereal grains in human diets. Over the past 10 years, wheat genetic technology for processing has advanced quickly. The initial genetic modification of cereals relied on the insertion of DNA into protoplasts and the subsequent development of calluses for the recovery of viable plants. To provide sustained output, a breeding program must maintain genetic diversity. To increase variation in genomes, plant breeders have widely used variation in genes from various gene pools. Therefore, it is necessary to search for an alternative, advanced, and affordable technique for genetically enhancing the gene pool and allelic diversification in order to get around the drawbacks of limited and homogenous genetic variations (Kazi *et al.* 2017). It is possible to significantly increase crop genetic diversity by combining tissue culture methods with plant biotechnology and breeding programs.

Many years ago, these techniques were used to modify variability as well as produce genetic diversity in order to enhance the genetic pool that was accessible and make it more acceptable for a plant breeder to employ for crop development. Cell protoplasts, anthers and microspores (immature pollen grains), ovaries and ovules, and embryos are all cultures of plants that contain genetic and epigenetic variability in the breeding stock. These in vitro culture techniques make use of all the genetic diversity that is now accessible and shorten the breeding program's time to create genotypes that are tolerant of and resistant to environmental stresses.

Plant tissue culture is mostly utilized in crops that are vegetatively propagated and self-pollinated, particularly those with smaller genetic bases. For instance, wheat, an autogamous crop, may have a limited genetic basis since due to the outcrossing reduces the possibility of genetic variation developing naturally to approximately 3%–5%. Therefore, enhancing the genetic base, retrieving acceptable variation, and controlling the desired feature all may be possible using in vitro approaches. The most common methods for genetic rearrangement and epigenetic reprogramming involve the induction of mutations, DNA and histone methylation, and histone acetylation. These methods include the application of physical and chemical mutagens, epigenetic agents like DNA demethylases and histone deacetylase inhibitors as well as in vitro methods. Likewise, the most promising method for genetic engineering desired features is through modern genome editing, together with plant tissue culture and *Agrobacterium* transformation. To enhance essential agronomic features, CRISPR/Cas9 (clustered regularly interspaced short palindromic repeats–associated protein 9) nuclease-mediated genome editing may precisely modify any gene or region of the plant genome. Traditional wheat breeding can accomplish the same thing, but it can take up to 7–10 years, whereas CRISPR technology yields results in a much shorter time (Kumar *et al.* 2019). In addition, the creation of somaclonal and gametoclonal variations makes plant tissue culture the most effective approach for crop development. With the isolation of important different variants in well-adapted, high-yield genotypes with greater disease resistance or stress tolerance capacities, the micropropagation technique offers an immense opportunity to create plants of superior quality (Brown and Thorpe 1995).

In this sense, one of the main difficulties for breeders is developing novel crop varieties that can withstand both abiotic and biotic stresses. Selection, hybridization, mutation induction via chemical and physical substances, and somaclonal variation have all been used in breeding techniques for decades (Table 1). More recently, wheat breeding has been notably made easier due to genome editing technology, availability of full genome sequences, effective tissue culture, and transformation procedures.

Historical Progress in Plant Tissue Culture

The discovery of the cell and the development of the cell theory serve as the foundation for the science of plant tissue culture. Schleiden and Schwann claimed that the cell is the fundamental structural component of all living things in 1838. They imagined that because each cell had the capacity for autonomy, they should all be able to regenerate into whole plants if given appropriate conditions. Based on this theory, a German physiologist named Gottlieb Haberlandt (1902) made an initial effort to cultivate isolated single palisade cells from leaves in a salt solution that had been enhanced with sucrose. The cells survived for up to a month, grew bigger, and gathered starch, but they did not divide. Despite his failure, he established the principles of tissue culture technology, for which he is credited as the father of plant tissue culture.

In order to create the first organ culture, White was the first person to successfully grow tomato roots in a synthetic nutritional media in 1934. In this study, tomato root tips might develop in a controlled laboratory environment. Later, researchers identified the chemical signals that enable plant tissues grown in vitro to differentiate (to produce a callus of parenchyma cells) and then dedifferentiate (to generate an embryo and plantlets). While Gautheret and Nobecaust were successful in creating callus cultures from cambial tissue in 1939. By cultivating and regenerating plants from meristems, Morel and Martin (1952) were able to successfully escape viral infection from the plants. In 1953, Skoog and Miller discovered the significance of the auxin and cytokinin ratio in various types of plant tissue culture after a year. Cocking then began isolating and cultivating protoplasts in 1960. As a result, in 1962, Murashige and Skoog developed a culture media for tobacco, which served as the foundation for the creation of a specified growth medium for different species. Carlson and his colleagues created somatic cell hybrids in 1972 by fusing the protoplasts of the two *Nicotiana* species. Now a days, genetic engineering has advantages from its use.

Tissue Culture: Present and Future Scenario

Over the past few decades, plant cell biotechnology has emerged as a new era in the discipline of biotechnology, concentrating on the production of a large number of secondary plant products. During the final decade of the last century, the development of genetic engineering and molecular biology techniques allowed the appearance of upgraded and new agricultural products that have appeared to fill an increasing demand in the productive system (Mastache 2007). However, without the invention of tissue culture methods, introducing genetic material into plant cells would not have been possible. The utilization of transgenic plants is currently one of the most promising approaches for manufacturing proteins and other therapeutic substances, such as antibodies and vaccinations (Ferrante and Simpson 2001). Transgenic crops are an affordable alternative to production techniques based on fermentation. Since plants are free of human diseases and therefore do not need to be screened for viruses and bacterial toxins, plant-made antibodies and vaccinations stand out in particular. In contrast to the 11 million farmers that did so in 2007, 13.3 million farmers used transgenic plants in their cultivation systems in 2008 (James 2008).

Application of Tissue Culture in Plant Breeding

Genetic Modification

The most recent development in plant cell and tissue culture is genetic alteration, which offers a method for transferring genes with desired traits into host plants and recovering transgenic plants (Hinchee *et al.* 1994). Incorporating the technique into plant biotechnology and breeding programs, it offers a significant potential for genetic improvement of diverse crop plants. It has the potential to play a significant role in the introduction of key agronomically significant characteristics like improved quality, increased yield, and improved resistance to diseases and pests. Either vector-mediated (indirect gene transfer) or vectorless (direct gene transfer) techniques may be used for genetic modifications in plants (Sasson 1993). The most popular approach for expressing foreign genes in plant cells that use a vector is *Agrobacterium*-mediated genetic transformation. The genetic transformation of root explants resulted in the successful introduction of agronomic features in plants. In order to produce recombinant proteins on a large scale, virus-based vectors provide an alternate method of stable and quick temporal protein expression in plant cells (Chung *et al.* 2006).

Direct transfer of DNA to mature seed-derived shoot apices using the particle bombardment method led to the spectacular development of transgenic *Jatropha* plants (Purkayastha *et al.* 2002). This method significantly reduces the amount of harmful compounds in seeds. Consequently, it overcomes the challenge of seed usage in numerous industrial sectors. It is now possible to regenerate plants that are resistant to viruses or disease by using genetic modification procedures. Researchers were able to create transgenic potato plants that are resistant to the potato virus Y, which poses a serious danger to potato crops all over the world (Bukovinszki *et al.* 2007).

Organogenesis

It is a process through which plant organs, such as roots, shoots, and leaves, are produced. These organs may develop directly from meristems or indirectly from calluses, which are undifferentiated cell masses. By adjusting the quantity of plant growth hormones in the nutritional media, plant regeneration via organogenesis includes the development of calluses and the differentiation of adventitious meristems into organs. Skoog and Miller were the first to show that a high cytokinin to auxin ratio encouraged root regeneration while a high auxin to cytokinin ratio increased the growth of shoots in tobacco callus.

Embryo Culture

The process of growing embryos from seeds and ovules in a nutritional media is known as embryo cultivation. In embryo cultivation, the plant either grows directly from the embryo or indirectly by developing a callus, which is followed by the development of shoots and roots. The method was created to thaw dormant seeds, assess the viability of seeds, and grow haploid plants and rare species (Holeman 2009). Growing excised embryos is a useful strategy used to minimize the breeding process of plants, which also reduces the duration of the long-dormant stage of seed. With the particular goal of mass multiplication, intravarietal hybrids of the commercially significant energy plant "*Jatropha*" have been successfully developed.

Apart from that, by using the embryo culture process, endangered species can be protected.

By removing embryos from mature seeds, an effective methodology has recently been designed for the in vitro multiplication of *Khaya grandifoliola* (Okere and Adegeye 2011).

Embryo Rescue

One of the oldest and most effective in vitro culture techniques is embryo rescue, which is used to aid in developing plant embryos that may not live to become surviving plants (Sage *et al.* 2010). The generation of numerous interspecific and intergeneric food and ornamental plant crop hybrids is made possible through embryo rescue, which is a crucial component of recent plant breeding. This method gives the young or weak embryo the care it needs to survive. Multicellular structures called plant embryos have the capacity to grow into new plants. Embryo culture is the most popular method of embryo rescue, which involves eliminating plant embryos and placing them on culture medium (Miyajuma 2006). In general, interspecific and intergeneric crosses that would otherwise result in seeds that are aborted are made through embryo rescue. Interspecific incompatibility in plants may happen for a variety of reasons, although embryo abortion is the most frequent one (Reed and Sandra 2005). Wide hybridization in plant breeding can produce tiny, shrunken seeds that show signs of fertilization but do not grow. Embryo rescue can help to solve this issue because remote hybridizations frequently fail to conduct typical sexual reproduction (Bridgen and Mark 1994).

Cell Suspension Culture

Today, techniques for cell suspension culture are utilized to culture plant cells on a massive scale so that secondary metabolites can be collected. The comparatively friable component of the callus is transferred into liquid media to create a suspension culture, which is then maintained under the proper conditions for aeration, agitation, light, temperature, and other physical properties (Chattopadhyay *et al.* 2002). Cell cultures are able to produce defined standard phytochemicals in significant quantities while also removing the presence of interference-causing substances seen in field-grown plants (Lila 2005). This approach has the benefit of eventually offering a consistent, credible source of natural products (Rao and Ravishankar 2002). According to Karuppusamy (2009), the main benefit of cell cultures is their ability to produce bioactive secondary metabolites in a controlled environment irrespective of soil and climate conditions.

Production of Haploids

By using anther, microspore cultures and protoplast, instead of traditional breeding, tissue culture techniques allow for the production of homozygous plants in a comparatively short period of time (Morrison and Evans 1998). By spontaneously or artificially inducing chromosome doubling, homozygous diploids, from haploid sterile plants with a single pair of chromosomes, can be created. Chromosome duplication restores the fertility of plants, producing double haploids that have the potential to develop into new pure breeding cultivars (Basu *et al.* 2011). The process of creating haploid plants from immature pollen cells without fertilization is known as androgenesis. By accelerating the creation of inbred lines and removing barriers like embryo nonviability and seed dormancy, haploidy technology has now become a crucial component of plant breeding programs (Bajaj 1990). The process produces haploid plants with induced tolerance to a variety of biotic and abiotic stresses, which is a notable application for the technology in genetic transformation.

The effective achievement of drought-tolerant plants was made possible by the insertion of genes for the desired feature during the haploid stage, followed by chromosomal doubling (Chauhan and Khurana 2011).

Production of Doubled Haploids

A significant development in wheat breeding is the generation of doubled haploids since it shortens the duration of breeding. Additionally, it is simpler to identify superior genotypes because doubled haploid lines have perfect homozygosity. The objectives of doubled haploid culture in plant breeding include the introduction of novel cultivars using the F1 double haploid system, the selection of mutants resistant to disease, the creation of asexual tree/ perennial species lines, and the introduction of desirable foreign genes.

Production of Virus-free Plants

Plant viral infections are easily contagious and reduce the plant's performance and output. Plant breeders are constantly interested in creating and cultivating virus-free plants since it is highly challenging to treat and heal plants that have been infected with viruses. Tissue culture has made it possible to commercially generate virus-free plants in some ornamental crops. This is accomplished by regenerating plants using meristems that are typically devoid of infection and cultured tissues obtained from virus-free plants. Because most viruses exist by creating a gradient in plant tissues, the size of the meristematic tissues utilized in cultures plays an important role in the virus's elimination.

Somaclonal Variation

In nature, recombination processes produce genetic variability and diversity within an individual population. Different factors, including natural selection, mutation, migration, and population size, have an impact on genetic variability. Higher plant cells cultivated in vitro displayed a genetic instability that was also a trait of regenerated cells, according to a 1958 study on a unique, intentionally produced source of genetic variability. It was reported to have made the first somaclonal variation observation. The variability present in plant tissue and cell cultures subsequently attracted a lot of interest, and Larkin and Scowcroft suggested neologisms to describe the outcomes of in vitro plant cultures.

The somaclone plants could differ physically from the existing donor plants to some extent or entirely. Variability typically arises spontaneously and may result from transient or long-term genetic alterations in cells or tissue during in vitro culture. Temporary alterations are nonheritable, reversible, and caused by physiological or epigenetic factors. Permanent modifications, on the other hand, are heritable and frequently exhibit preexisting plant variability in the original plant or are the consequence of de novo variation. According to the available literature, somaclonal variation can affect anything from a single characteristic to the entire plant genome. Improvement is the main advantage of somaclonal variation in plants. Additional genetic variability is produced as a result of somaclonal variation. Resistance to disease-causing pathogens, herbicide or chemical tolerance, environmental stress tolerance, and enhanced output of secondary metabolites are some of the traits that somaclonal mutants can be enriched for during in vitro culture. Irrespective of the seasons or the plants, micropropagation can be done throughout the entire year. Somaclonal variation can be seen in processes which demand for clonal uniformity, such as in the horticulture and forestry sectors where tissue culture is used to quickly propagate elite genotypes.

Protoplast Fusion

The process of fusing protoplasts with genomes from two distinct species involves selecting the necessary somatic hybrid cells and then regenerating hybrid plants. The ability to transmit desired traits via protoplast fusion from one species to another is becoming increasingly important for crop development (Brown and Thorpe 1995). By electrofusing rice and ditch reed protoplasts for salt tolerance, somatic hybrids were created (Mostageer and Elshihy 2003). By removing the constraints of sexual incompatibility, in vitro fusing of protoplasts creates a path for the development of specific hybrid plants.

The method has been used in the horticultural sector to develop new hybrids that produce more fruit and remain more disease-resistant. When citrus protoplasts were fused with other closely related citrinae species, successful viable hybrid plants were produced (Motomural *et al.* 1997). A successful protocol has been devised for the development of somatic hybrid plants employing two different types of wheat protoplast as the recipient and protoplast of *Haynaldia villosa* as the fusion donor in order to address the issue of chromosomal loss and poor regeneration potential. With the aim to improve wheat, it is also used as a significant gene source (Liu *et al.* 1988) (Flowchart 1).

Flowchart 1: Schematic representation of production of hybrid plant via protoplast fusion.

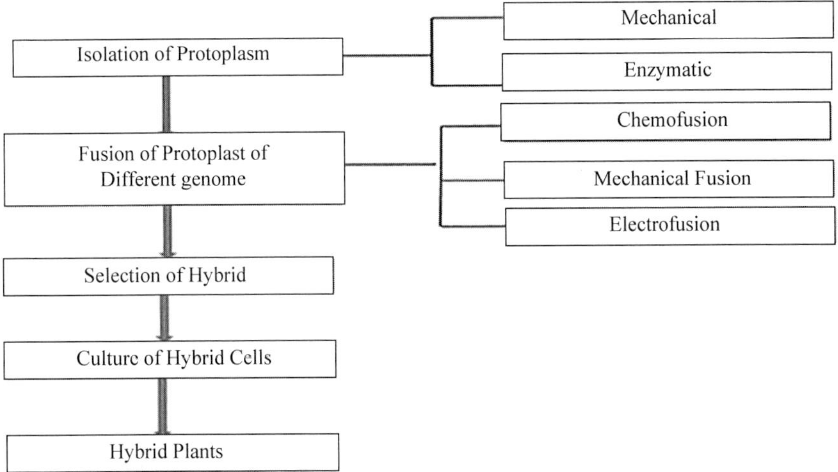

Production of Artificial Seeds

The somatic embryos are enclosed in synthetic seeds in an appropriate matrix (such as sodium alginate) as well as insecticides, fungicides, mycorrhizae, and herbicides. These synthetic seeds can be used to quickly and extensively reproduce desired plant species as well as hybrid genotypes.

Synthetic seeds have several advantages over natural seeds, including the capacity to be stored for up to a year without losing viability, ease of handling and utility as delivery units, the ability to be put directly in the ground like real seeds, and the lack of requirement for greenhouse acclimatization.

Conclusion

An essential component of applied biotechnology is tissue culture. Global climate change is another factor to take into account as the population of the world grows and living quarters and agricultural grounds are reduced drastically. We must make sure that the planet for our future generation is calm, healthy, and free from hunger. There is no substitute for plant tissue culture for this. Plant tissue culture is currently a well-recognized technology that has greatly aided in the expansion and development of agricultural products.

The tissue culture approach is currently used in plant breeding. It offers the means of genetic modification, somatic embryogenesis and organogenesis, embryo rescue, haploid and double haploid production, production of the virus-free plants, protoplast fusion in production of the required hybrids, somaclonal variation to the creation of additional genetic variability, micropropagation, production of synthetic seed, and cryopreserved for further application. Plant tissue culture generally represents the most potential areas of application at the moment with opportunities ahead.

References

Bajaj, Y.P.S. 1990. "In Vitro Production of Haploids and Their Use in Cell Genetics and Plant Breeding" In Haploids in Crop Improvement I Berlin: Springer Berlin Heidelberg. pp. 3-44.

Bridgen, M., Van Houtven W., Eeckhaut T. 2018. "Plant Tissue Culture Techniques for Breeding." In Ornamental Crops: Handbook of Plant Breeding edited by Johan Van Huylenbroeck. Cham, Switzerland: Springer International Publishing. pp. 127-44.

Brown, D.C., Thorpe T.A. "Crop Improvement through Tissue Culture." World Journal of Microbiology and Biotechnology 11, 4 (1995):409-15.

Bukovinszki, A,, Divéki Z., Csányi M., Palkovics L., Balázs E. "Engineering Resistance to PVY in Different Potato Cultivars in a Marker-free Transformation System Using a 'Shooter Mutant' A. tumefaciens." Plant Cell Reports 26, 4 (2007):459-65.

Chauhan, H., Khurana P. "Use of Doubled Haploid Technology for Development of Stable Drought Tolerant Bread Wheat (Triticum aestivum L.) Transgenics." Plant Biotechnology Journal 9, 3 (2011):408-17.

Chung S.M., Vaidya M., Tzfira T. "Agrobacterium is not Alone: Gene Transfer to Plants by Viruses and Other Bacteria." Trends in Plant Science 11, 1 (2006):1-4.

Ferrante, E., Simpson D. "A Review of the Progression of Transgenic Plants Used to Produce Plantibodies for Human Usage." Journal of Young Investigators 4, 1 (2001): 11.

Grassini, P., Eskridge K.M., Cassman K.G. "Distinguishing between Yield Advances and Yield Plateaus in Historical Crop Production Trends." Nature Commununications 4 (2013): 2918.

Holeman, D.J. 2009. Simple Embryo Culture for Plant Breeders. A Manual of Technique for the Extraction and In Vitro Germination of Mature Plant Embryos with Emphasis on the Rose, 1st ed. https://rosebreeders.org/embryoculture.pdf [Last accessed December, 2023].

James, C. 2008. Global Status of Commercialized Biotech/GM Crops: 2008. ISAAA Brief No. 37 Ithaca, NY: International Service for the Acquisition of Agri-biotech Applications.

Kazi, A.M., Ali N., Ibrahim A., Napar A.A., Jamil M., Hussain S., et al. "Tissue Culture Mediated Allelic Diversification and Genomic Enrichment of Wheat to Combat Production Constraints and Address Food Security." Plant Tissue Culture and Biotechnology 27 (2017):89-140.

Kumar, R., Kaur A., Pandey A., Mamrutha H.M., Singh G.P. "CRISPR-based Genome Editing in Wheat: A Comprehensive Review and Future Prospects." Molecular Biology Reports 46 (2019):3557-69.

Liu, J.D. 1988. "Transfer of Haynaldia villosa Chromosomes into Triticum aestivum" Proceedings of the Seventh International Wheat Genetics Symposium edited by TE Miller and RMD Koebner. Cambridge: Institute of Plant Science Research. pp. 355-61.

Mastache, L.C.N. "Large Scale Commercial Micropropagation in Mexico. The Experience of Agromod, SA de CV." Acta Horticulturae 748 (2007):91-4.

Morrison, R.A., Evans D.A. "Haploid Plants from Tissue Culture: New Plant Varieties in a Shortened Time Frame." Bio/Technology 6, 6 (1988):684-90.

Mostageer, A., Elshihy O.M. "Establishment of a Salt Tolerant Somatic Hybrid through Protoplast Fusion between Rice and Ditch Reed." Arab Journal of Biotechnology 6, 1 (2003):1-12.

Motomural, T., Hidaka T., Akihamal T., Katagi S., Berhow M., Moriguchi T., et al. "Protoplast Fusion for Production of Hybrid Plants between Citrus and its Related Genera." Journal of the Japanese Society for Horticultural Science 65,4 (1997):685-92.

Okere, A.U., Adegeye A. "In Vitro Propagation of An Endangered Medicinal Timber Species Khaya grandifoliola C. Dc." African Journal of Biotechnology 10, 17 (2011):3335-9.

Ray, D.K., Mueller N.D., West P.C., Foley J.A. "Yield Trends are Insufficient to Double Global Crop Production by 2050." PLoS ONE 8 (2013): e66428.

Salina, E.A., Adonina I.G., Badaeva E.D., Kroupin P.Y., Stasyuk A.I., Leonova I.N., et al. "A Thinopyrum intermedium Chromosome in Bread Wheat Cultivars as a Source of Genes Conferring Resistance to Fungal Diseases." Euphytica 204 (2015):91-101.

Sasson., A. 1993. Biotechnologies in Developing Countries: Presmtami Future, vol. 1. Paris: United Nations Educational, Scientific and Cultural Organization.

Slama O.A., Amara H.S. 2021. "Unpollinated Ovaries Used to Produce Doubled Haploid Lines in Durum Wheat" Doubled Haploid Technology. Germany: Springer. pp. 245-55.

7

Breeding Approaches in Transgenic Crops: Ethical Issues and Regulation

Deepa Bhadana [1]***, Sajalsaha*** [2]***, Rinkey Arya*** [3] ***and Rajib Das*** [4]

[1]*Ph.D. Chaudhary Charan Singh University, Meerut, Uttar Pradesh, India*

[2]*Ph.D. Nagaland University, School of Agricultural Sciences, Medziphema, Nagaland India*

[3]*Faculty of Agriculture and Agroforestry, DSB Campus Nainital, Kumaun University Uttarakhand, India*

[4]*Assistant Professor, College of Agriculture, Central Agricultural University, Pasighat Arunachal Pradesh, India*

Abstract

Transgenic crops, developed through genetic engineering techniques, have revolutionized modern agriculture by conferring novel traits that enhance yield, quality, and resilience. This chapter delves into the intricate interplay between the scientific advancements in breeding approaches for transgenic crops and the critical ethical concerns and regulatory frameworks that accompany them. The chapter begins by introducing the concept of transgenic crops and provides an overview of the diverse breeding techniques employed to manipulate plant genomes. These techniques include Agrobacterium-mediated transformation, biolistic transformation, CRISPR-Cas9 (clustered regularly interspaced short palindromic repeats–associated protein 9) genome editing, RNA interference (RNAi), and site-specific integration methods. Each method is presented within the context of its potential benefits and limitations. The subsequent sections address the ethical considerations that arise from the creation and deployment of transgenic crops. The potential ecological and biodiversity impacts, cross-contamination risks with conventional crops, and issues related to patenting genetically modified organisms (GMOs) are explored. Furthermore, the chapter examines the ethical implications of genetically modified crops on human and animal health and questions surrounding the monopolization of seed resources. Looking to the future, the chapter outlines emerging breeding technologies and anticipates potential challenges in ethical consideration and regulation. It concludes by emphasizing the necessity of a balanced approach that harnesses the benefits of transgenic crops while respecting ethical principles and ensuring effective regulatory oversight.

Keywords: *Transgenic breeding, genetic engineering, genome, RNA interference*

Introduction

Introduction to Transgenic Crops and Breeding Approaches

Transgenic crops, at the forefront of modern biotechnology, represent a pivotal innovation in the field of agriculture. These crops, also known as genetically modified crops, are the result of intentional genetic manipulation, where specific genes from one organism are inserted into the genome of another, often leading to the acquisition of desired traits. This chapter introduces the fascinating world of transgenic crops by delving into their core definition and the array of breeding approaches harnessed to engineer these transformative organisms.

Definition and Explanation of Transgenic Crops

Transgenic crops are plants that have been genetically modified using advanced molecular techniques to introduce specific genes from different species. The genetic modifications are precisely tailored to confer advantageous traits, such as resistance to pests, diseases, and adverse environmental conditions as well as improved nutritional content and enhanced yield potential. These genetic manipulations enable plants to express proteins that provide them with novel capabilities not naturally present in their genomes. By seamlessly incorporating genes from unrelated organisms, transgenic crops break down the barriers that previously confined traditional breeding methods. This enables the transfer of genes across species boundaries, resulting in crops with traits that were previously unattainable through conventional breeding.

Breeding Approaches in Transgenic Crops

- ***Genetic engineering techniques***: Genetic engineering techniques are at the heart of transgenic crop development, allowing scientists to manipulate the genetic makeup of organisms for specific purposes. This section provides a comprehensive overview of the key genetic engineering techniques used in creating transgenic crops, highlighting their strengths, limitations, and comparative features.
- ***Agrobacterium-mediated transformation***: *Agrobacterium tumefaciens*, a soil bacterium, is harnessed as a natural genetic engineer. It transfers a portion of its DNA, known as a T-DNA, into plant cells. Researchers modify the T-DNA to carry desired genes, which are then incorporated into the plant's genome. This technique is well-suited for dicot plants

and is valued for its relatively high efficiency and ability to integrate genes at specific locations (Fig. 1) Pacurar *et al.*, 2011.

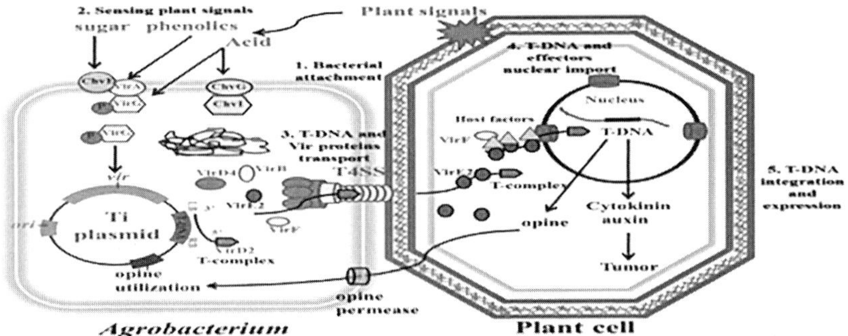

Fig. 1: Working system of *Agrobacterium* in plant cell.

Source: Hwang *et al.* 2017.

- ***Biolistic transformation (gene gun)***: Biolistic transformation involves using a "gene gun" to propel microscopic particles coated with genetic material into plant cells. The particles penetrate cell walls and membranes, delivering genes to the plant's DNA. This method is versatile and applicable to a wide range of plant species, including monocots and dicots. It enables direct insertion of genes, making it particularly useful when the target organism is recalcitrant to other techniques (Fig. 2).

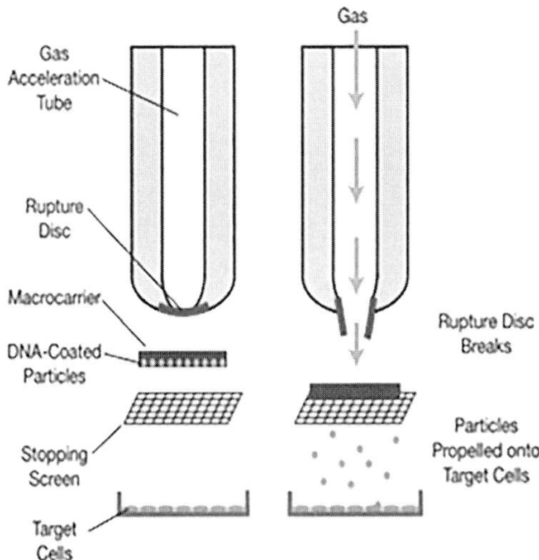

Fig. 2: Particle *bombardment* method of gene transfer.

Source: Demirer 2017.

- *CRISPR-Cas9 and genome editing*: Clustered regularly interspaced short palindromic repeats (CRISPR) and the Cas9 protein form a revolutionary genome editing system. It allows precise modification of specific DNA sequences. Researchers design a guide RNA that directs Cas9 to the desired target, where it introduces changes like insertions, deletions, or substitutions. CRISPR-Cas9 offers exceptional precision and has transformed genetic editing due to its simplicity and adaptability (Fig. 3) Khatodia *et al.* 2016.

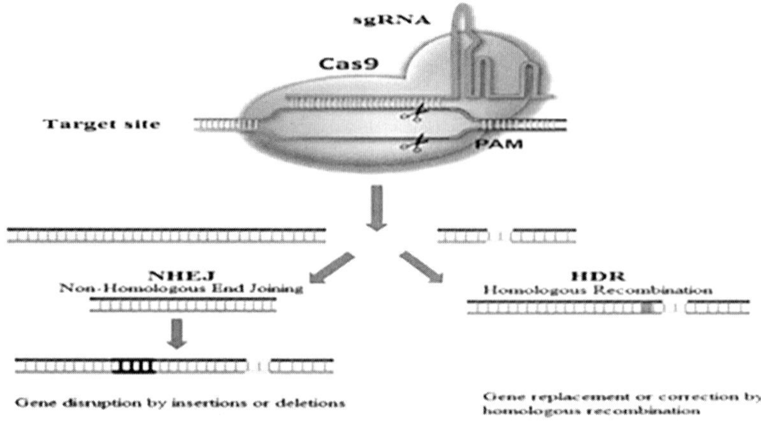

Fig. 3: Working system of CRISPR-Cas9. (CRISPR-Cas9: clustered regularly interspaced short palindromic repeats–associated protein 9; PAM: protospacer adjacent motif; sgRNA: single guide RNA)

Source: Tavakoli *et al.* 2021.

- *RNA interference (RNAi) technology*: RNAi involves introducing small RNA molecules that interfere with the expression of specific genes. This technique enables the silencing of target genes, either to suppress undesirable traits or enhance beneficial traits. RNAi is particularly effective against pests and pathogens, making it a valuable tool for crop protection (Fig. 4).

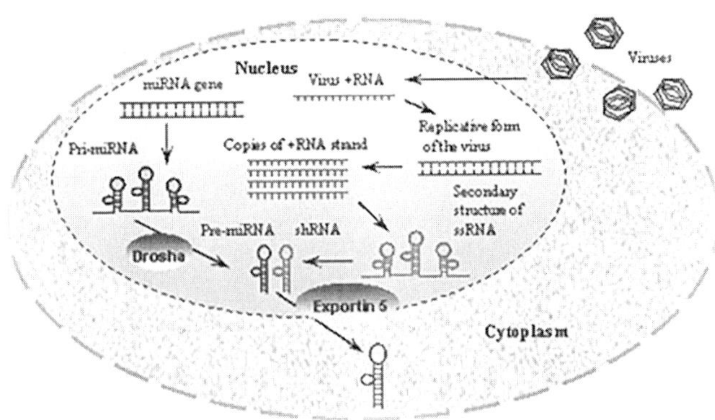

Fig. 4: Working system of RNA interference technology. (miRNA: microRNA; shRNA: small hairpin RNA; ssRNA: single-stranded RNA)

Source: RNA Interference (RNAi). https://www.ncbi.nlm.nih.gov/probe/docs/techrnai/ [Last accessed December, 2023].

- ***Site-specific integration methods***: Site-specific integration techniques focus on precisely inserting genes into predetermined locations within the genome. They minimize the risk of disrupting essential genes and optimize the stability of introduced traits. Recombinases, enzymes that facilitate DNA rearrangements, are often employed to achieve controlled gene insertion https://agrimoon.com/wp-content/uploads/Principles-of-Plant-Biotechnology.pdf.
- ***Comparative analysis***: Each genetic engineering technique has its advantages and considerations. *Agrobacterium*-mediated transformation is esteemed for its targeted integration and efficiency with dicotyledonous plants. Biolistic transformation offers versatility but may result in random integration. CRISPR-Cas9 excels in precise editing but might have off-target effects. RNAi technology is highly effective for gene silencing, while site-specific integration methods enhance predictability.

Benefits of Transgenic Crops

- ***Improved yield and crop quality***: Transgenic crops can be engineered to exhibit enhanced yield potential and improved crop quality. By introducing genes that regulate growth, flowering, and fruiting, these crops can produce higher yields while maintaining desirable characteristics such as size, color, and taste.
- ***Enhanced nutritional content***: Genetic engineering allows for the enrichment of crop nutritional content. For example, "*golden rice*"

has been developed to produce higher levels of vitamin A, addressing deficiencies that can lead to blindness and other health issues in regions where rice is a staple food.

- ***Pest and disease resistance*:** Transgenic crops can express proteins that confer resistance to pests and diseases. For instance, the incorporation of genes from the bacterium *Bacillus thuringiensis* results in crops that produce toxins toxic to specific insects, reducing the need for chemical pesticides.
- ***Environmental sustainability*:** The reduced dependence on chemical pesticides and the increased resistance to pests can lead to more environmentally sustainable agricultural practices. This benefits ecosystems by decreasing the negative impact of pesticide runoff and minimizing harm to nontarget organisms.
- ***Reduced chemical pesticide usage*:** Through pest-resistant transgenic crops, farmers can substantially reduce the use of chemical pesticides. This not only lowers production costs but also diminishes potential risks to human health, wildlife, and the environment.
- ***Adaptation to challenging environments*:** Transgenic crops can be engineered to thrive in challenging environments, such as regions with limited water availability or high salinity. These traits can contribute to improved food security in areas prone to climate-related stresses.
- ***Extended shelf life*:** Genetic modifications can enhance the shelf life of harvested crops by slowing down ripening processes or reducing susceptibility to spoilage. This can minimize post-harvest losses and enhance food availability.
- ***Reduced soil erosion*:** Certain transgenic crops, such as those with improved root systems, can help prevent soil erosion. Stronger roots stabilize the soil, preventing erosion caused by wind and water.
- ***Biofortification for nutrient-rich crops*:** Beyond vitamin enrichment, transgenic crops can be designed to accumulate essential minerals like iron and zinc in their edible parts. This addresses nutrient deficiencies prevalent in diets of vulnerable populations.
- ***Economic benefits for farmers*:** The increased yield and reduced inputs associated with transgenic crops can lead to economic gains for farmers. Moreover, by addressing yield losses due to pests and diseases, farmers are better equipped to manage risks and maintain consistent productivity.

- ***Innovation and research advancement*:** The development of transgenic crops stimulates scientific research and technological innovation. It encourages interdisciplinary collaboration and the exploration of cutting-edge solutions to agricultural challenges.

Achievements

- ***Transfer of genes for herbicide tolerance*:** Genes conferring herbicide tolerance have been successfully incorporated into plants, notably through the work of Shah and colleagues in 1986. They isolated a glyphosate-tolerant gene in *Petunia hybrida*, which was then introduced into nontolerant plants using the cauliflower mosaic virus (CaMV) 35 promoter. The resulting transgenic plants displayed glyphosate tolerance, while control plants perished when exposed to the herbicide. In 1987, De Block and his team transferred the bar gene from *Streptomyces hygroscopicus*, providing resistance to bialaphos and phosphinothricin, to tobacco, tomato, and potato using the 35S promoter from CaMV. This gene encodes an enzyme called phosphinothricin acetyltransferase, counteracting herbicide toxicity. This approach has revolutionized crop development, enabling the cultivation of herbicide-resistant crops with enhanced agricultural capabilities.
- ***Expression of insect tolerance in transgenic plants*:** *B. thuringiensis* is a bacterium known for producing proteinaceous crystals during sporulation, which possess insecticidal properties, particularly effective against lepidopteran insects. Utilizing *B. thuringiensis* as a microbial insecticide offers several advantages compared to chemical control agents. Its insecticidal crystal proteins (ICPs) exhibit species-specific action, rendering it harmless to nontarget insects, vertebrates, the environment, and users. In 1987, Fischhoff and his team engineered chimeric genes that combined the CaMV 35S promoter with coding sequences for *B. thuringiensis* crystal proteins. These genetically modified *B. thuringiensis* genes were successfully introduced into tomato and tobacco plants. Consequently, the transgenic plants exhibited heightened resistance to lepidopteran insects, providing an environmentally friendly and targeted approach to pest control.
- ***Expression of coat protein (CP) genes for virus protection*:** Abel and his colleagues conducted an experiment in which they introduced a chimeric gene into tobacco cells. This gene contained a cloned cDNA of the CP gene from TMV (tobacco mosaic virus). They utilized Ti plasmid from *A. tumefaciens*, which had its tumor-inducing genes removed, as the

vehicle for gene transfer. Plants that regenerated from these transformed cells exhibited the expression of TMV mRNA (messenger RNA) and CP as a nuclear trait. To assess the effects, seedlings from self-fertilized transgenic plants were exposed to TMV and closely monitored for the development of disease symptoms. Interestingly, the seedlings that carried the CP gene demonstrated a delay in the onset of symptoms, with a notable percentage (ranging from 10% to 60%) of the transgenic plants showing resistance and failing to develop any symptoms caused by TMV infection. This research highlights the potential for genetic modification to confer resistance to viral diseases in plants.

- ***Expression of antisense RNA in transgenic plants*:** Antisense RNA is a naturally occurring mechanism found in various organisms that serves to regulate gene expression. It exerts its inhibitory effects on gene expression by preventing ribosome binding, interfering with mRNA transport from the nucleus, and increasing the degradation of mRNA. In 1987, Rothstein and his team demonstrated the effectiveness of inhibiting the expression of the nopaline synthase (NOS) gene in tobacco using antisense RNA. They achieved this by introducing an NOS antisense gene construct driven by the CaMV 35S promoter into transgenic plants carrying the NOS gene. The resulting transformed plants were analyzed for NOS enzyme activity, which were found to vary depending on the tissue examined. This mechanism holds promise as a valuable tool in genetic modification, particularly when plants are engineered with antisense genes to suppress the expression of various undesirable traits or characteristics.

Points to be Considered in all Gene Transfer Programs

- ***Critical need for effective cell selection*:** Robust systems for selecting transformed cells become indispensable when dealing with more complex traits.
- ***Positional influence on gene expression*:** The expression of a specific gene within the transformed cell is contingent upon its location in the host genome. This underscores the necessity for further investigations into the positional effects of introduced genes and their subsequent expression.
- ***Challenges in regenerating diverse plants*:** While efficient gene transfer methods are available, the primary remaining hurdle lies in the limited range of plant species that can be successfully regenerated from transformable cells. Thus, the development of efficient regeneration systems tailored to specific crop species is imperative for successful gene transfer.

- ***Understanding regulatory mechanisms*:** A deeper understanding of the regulatory mechanisms involved in supplying essential substances required for the expression of specific genes is essential. This involves exploring whether the introduction of a new gene automatically enhances the supply of necessary substances or if other genes involved in substance synthesis need to be amplified.
- ***Mechanisms governing improved gene expression*:** Detailed investigations into the mechanisms responsible for regulating substances crucial for enhanced gene expression are essential for optimizing the gene transfer process.
- ***Metabolic alterations with new genes*:** Introducing novel genes into plant cells may necessitate the presence of new enzymes not originally found in these cells. Consequently, comprehensive studies into the alterations in the plant's entire metabolism are warranted.
- ***Stability of alien gene expression*:** Understanding the potential for the loss of expression of foreign genes over time is crucial. Research should be conducted to assess the stability of gene expression and its inheritance.
- ***Molecular analysis of target genes*:** Prior to embarking on gene transfer, a thorough exploration of the molecular aspects of the target gene is imperative. This evaluation helps ascertain the necessity and desirability of the gene in question.
- ***Challenges in stable gene expression*:** Delving into the challenges associated with maintaining the stable expression of foreign genes in crop plants is essential. This exploration can aid in developing strategies to modulate the temporal and spatial expression of specific genes.

Ethical Issues in Transgenic Crop Breeding

Transgenic crop breeding, while offering promising benefits, also raises a range of ethical concerns that must be carefully considered. This section explores these ethical issues in detail:

- ***Potential for unintended consequences in ecosystems*:** The release of transgenic crops into the environment carries the risk of unintended ecological consequences. Genes introduced into one species might transfer to other organisms through hybridization or horizontal gene transfer, potentially affecting nontarget species and disrupting delicate ecological balances.
- ***Cross-contamination with non-genetically modified organisms (GMOs)*:** Transgenic crops have the potential to crossbreed with

nongenetically modified crops or wild relatives, resulting in the spread of transgenes beyond intended fields. This raises concerns about the creation of hybrid plants with unpredictable traits, which could have ecological, economic, and societal repercussions.

- ***Patenting of GMOs*:** The patenting of transgenic crops and their genes raises ethical concerns about ownership of life forms. Patenting can create legal barriers to using and further developing certain crop varieties, and potentially limiting farmers' and researchers' access to critical genetic resources.
- ***Impact on biodiversity*:** The widespread adoption of a few transgenic crop varieties with specific traits could reduce the genetic diversity of crops. This can make entire agricultural systems more vulnerable to pests, diseases, and changing environmental conditions, potentially endangering global food security.
- ***Ethical concerns related to animal and human health*:** The consumption of transgenic crops raises concerns about potential risks to human health. Although rigorous testing is conducted, there may still be unknown long-term effects on human health, which raises ethical questions about exposing populations to potential risks. Additionally, concerns arise about feeding transgenic crops to animals and the possible consequences for animal health and well-being.

Regulatory Framework for Transgenic Crops

In India, all activities related to genetically engineered (GE) organisms or cells and hazardous microorganisms and products thereof are regulated as per the "Manufacture, Use, Import, Export and Storage of Hazardous Microorganisms/ Genetically Engineered Organisms or Cells, Rules, 1989" (Rules, 1989) notified by the Ministry of Environment, Forest and Climate Change (MoEF and CC), Government of India under the Environment (Protection) Act, 1986 (EPA 1986).

Definitions of Genome-edited Plants derived from Use of Site-directed Nucleases

The genome-edited plants derived from the use of genome editing techniques employing site-directed nucleases (SDNs) such ZFNs (zinc-finger nucleases), TALENs (transcription activator-like effector nucleases), CRISPR, and other nucleases with similar functions are generally classified under three categories as: (1) SDN-1, a site-directed mutagenesis without using a DNA sequence template; (2) SDN-2, a site-directed mutagenesis using a DNA sequence template; and (3) SDN-3, site-directed insertion of gene/large DNA sequence using a DNA sequence template (Flowchart 1).

Flowchart 1: Process of regulation of genome editing plants. (BRL: biosafety research level; IBKP: Indian Biosafety Knowledge Portal; IBSC: Institutional BioSafety Committee; RCGM: Review Committee on Genetic Manipulation; SDN: site-directed nuclease)

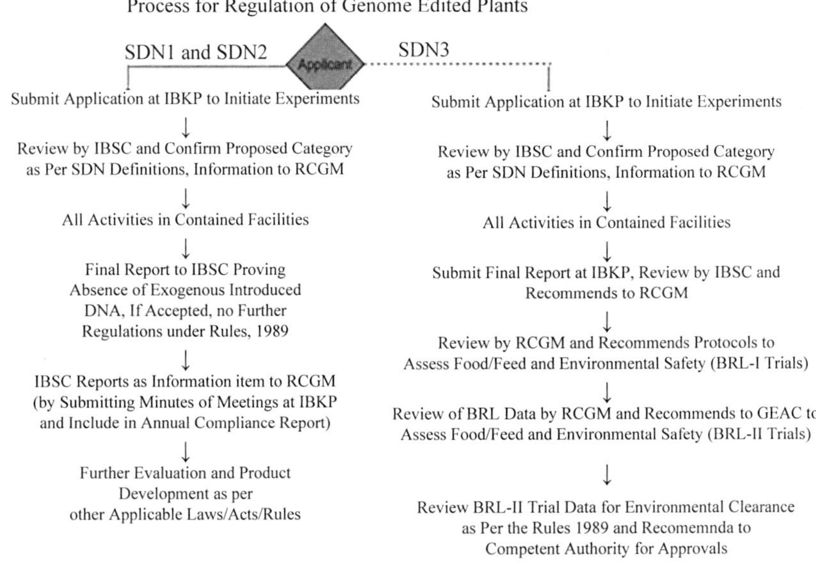

Six Competent Authorities, their Mandate, and Functions as Defined Under Rules 1989

1. ***Recombinant DNA Advisory Committee (RDAC)***: The RDAC functions in Department of Biotechnology (DBT) and shall review developments in biotechnology at national and international levels and shall recommend suitable and appropriate safety regulations for India in recombinant research, use, and applications from time to time.
2. ***Institutional Biosafety Committee (IBSC)***: The IBSC is constituted by an occupier or any person including research institutions, handling hazardous microorganisms/genetic engineered organisms (at research and development level). The occupier or any person including research institutions shall prepare, with the assistance of the IBSC, an up-to-date on-site emergency plan according to the manuals/guidelines of the RCGM (Review Committee on Genetic Manipulation) and make available copies to the DLC/SBCC (District Level Committee/State Biotechnology Coordination Committee) and RCGM/GEAC (Genetic Engineering Appraisal Committee).
3. ***RCGM***: RCGM functions in the DBT to monitor the safety-related aspects in respect of ongoing research projects and activities involving

GE organisms/hazardous microorganisms. It shall bring out manuals of guidelines specifying procedure for regulatory process with respect to activities involving GE organisms in research, use, and applications with a view to ensure environmental safety.

4. ***GEAC*:** The GEAC functions in the MOEF and CC and is responsible for approval of (1) activities involving large-scale use of hazardous microorganisms and recombinants in research and industrial production from environmental angle and (2) proposals relating to release of GE organisms and products into the environment including experimental field trials.
5. ***SBCC*:** SBCC reviews periodically the safety and control measures in the various installations/institutions handling GE organisms/hazardous microorganisms. SBCC has powers to inspect, investigate, and take punitive action in case of violations of statutory provisions through the Nodal Department and the State Pollution Control Board/Directorate of Health/Medical Services.
6. ***DLC*:** DLC monitor the safety regulations in installations/institutions engaged in the use of GMOs/ hazardous microorganisms and its applications in the environment. DLC shall also prepare an off-site emergency plan for field trials. DLC shall regularly submit its report to the SBCC/GEAC.

Future Directions and Challenges

- ***Emerging technologies in transgenic crop breeding*:** The landscape of transgenic crop breeding is continually reshaped by emerging technologies. Techniques like CRISPR-Cas12 and CRISPR-Cas13 offer increased precision and capabilities for genome editing beyond Cas9. Other advancements, such as synthetic biology and genome synthesis, hold the potential to engineer entirely novel traits in crops.
- ***Addressing ethical concerns through innovative solutions*:** The ethical concerns surrounding transgenic crops require innovative solutions. Researchers are exploring ways to minimize unintended ecological consequences through gene containment methods, such as gene drives that spread engineered traits through populations. Additionally, efforts are underway to enhance the nutritional content of crops in ways that resonate with ethical considerations.
- ***Navigating evolving regulatory landscapes*:** The rapid pace of technological innovation presents challenges to regulatory frameworks. As new genetic engineering techniques emerge, regulators must grapple

with defining and classifying organisms, deciding which falls under regulatory oversight. Ensuring these frameworks are flexible enough to accommodate innovation while maintaining safety is a critical challenge.

- ***Consumer acceptance and communication***: Bridging the gap between scientific advancements and public perception remains a challenge. Effective communication strategies are essential to convey the benefits, risks, and ethical considerations associated with transgenic crops. Building public trust and addressing misconceptions require continuous efforts from scientists, policymakers, and communicators.
- ***Sustainability and long-term impacts***: While transgenic crops can address short-term challenges, it is crucial to consider their long-term sustainability. Evaluating the impact of transgenic crops on soil health, biodiversity, and ecosystems over extended periods is essential to ensure that the benefits outweigh potential negative consequences.
- ***Global access and equity***: Ensuring equitable access to transgenic crop technologies is paramount. This involves addressing issues related to intellectual property rights, licensing, and affordability, especially for resource-constrained regions that could benefit significantly from these innovations.
- ***Collaboration and multidisciplinary research***: Tackling the complex challenges of transgenic crop breeding requires collaboration among scientists, ethicists, policymakers, economists, and various stakeholders. Multidisciplinary research can help anticipate and address potential issues holistically.
- ***Anticipating unintended consequences***: As new traits are engineered into crops, it is crucial to carefully anticipate and assess potential unintended consequences that might only become evident over time. Continuous monitoring and research are necessary to catch and mitigate any unforeseen impacts.

Conclusion

In this chapter, we have explored the multifaceted landscape of transgenic crop breeding, delving into the scientific advancements, ethical considerations, and regulatory frameworks that shape this evolving field. Key points discussed throughout the chapter underscore the complexity and importance of transgenic crop development. Transgenic crops, also known as GMOs, are the result of intentional genetic manipulation, introducing novel traits into plant genomes through techniques such as Agrobacterium-mediated transformation, gene gun delivery, CRISPR-Cas9 genome editing, RNAi, and site-specific integration methods. These crops offer a range of benefits, including improved yield and crop quality, enhanced nutritional content, resistance to pests and diseases, reduced reliance on chemical pesticides, and adaptability to challenging environments. However, along with these benefits come ethical concerns that require careful consideration. Ethical issues in transgenic crop breeding encompass potential unintended ecological consequences, cross-contamination with non-GMO organisms, patenting of GMOs, impact on biodiversity, and concerns about animal and human health. The regulatory framework for transgenic crops involves international agreements, government agency oversight, biosafety assessments, risk management strategies, and requirements for labeling and traceability. These mechanisms aim to ensure the safety of transgenic crops for human health, the environment, and agricultural practices. As we look to the future, emerging technologies like advanced genome editing techniques and synthetic biology hold the potential to reshape transgenic crop breeding. Addressing ethical concerns will require innovative solutions, such as gene containment methods and enhanced nutritional biofortification. Navigating evolving regulatory landscapes will be crucial, especially as new genetic engineering techniques challenge existing definitions and oversight protocols. Effective communication with the public and collaborative efforts between stakeholders will play essential roles in shaping public perception and acceptance of transgenic crops.

References

De Block, M., Botterman J., Vandewiele M., Dockx J., Thoen C., Gosselé V., et al. "Engineering Herbicide Resistance in Plants by Expression of a Detoxifying Enzyme." The EMBO Journal 6, 9 (1987):2513-8.

Demirer G.S., Landry M.P. 2017. Delivering Genes to Plants. 113(4). https://www.aiche.org/resources/publications/cep/2017/april/delivering-genes-plants [Last accessed December, 2023].

Fischhoff, D.A., Bowdish K.S., Perlak F.J., Marrone P.G., McCormick S.M., Niedermeyer J.G., et al. "Insect Tolerant Transgenic Tomato Plants." Bio/Technology 5, 8 (1987):807-13.

Hwang, H.H., Yu M., Lai E.M. "Agrobacterium-mediated Plant Transformation: Biology and Applications" Arabidopsis Book 15 (2017): e0186.

Khatodia, S., Bhatotia K., Passricha N., Khurana S.M.P., Tuteja N. "The CRISPR/Cas Genome-Editing Tool: Application in Improvement of Crops." Frontiers in Plant Science 7 (2016): 506.

Pacurar, D.I., Thordal-Christensen H., Pacurar M., Pamfil D.C. "Agrobacterium tumefaciens: From Crown Gall Tumors to Genetic Transformation." Physiological and Molecular Plant Pathology 76, 2 (2011):76-81.

Principles of Plant Biotechnology. ICAR eCourse. https://agrimoon.com/wp-content/uploads/Principles-of-Plant-Biotechnology.pdf [Last accessed December, 2023].

RNA Interference (RNAi). https://www.ncbi.nlm.nih.gov/probe/docs/techrnai/ [Last accessed December, 2023].

Rothstein, S.J., DiMaio J., Strand M., Rice D. "Stable and Heritable Inhibition of the Expression of Nopaline Synthase in Tobacco Expressing Antisense RNA." Proceedings of the National Academy of Sciences 84, 23 (1987):8439-43.

Shah D.M., Horsch R.B., Klee H.J., Kishore G.M., Winter J.A., Tumer N.E., et al. "Engineering Herbicide Tolerance in Transgenic Plants." Science 233, 4762 (1986):478-81.

Tavakoli, K., Pour-Aboughadareh A., Kianersi F., Poczai P., Etminan A., Shooshtari L. "Applications of CRISPR-Cas9 as an Advanced Genome Editing System in Life Sciences." BioTech (Basel (Switzerland)) 10, 3 (2021): 14.

8

Marker-assisted Selection: Approaches and Importance in Plant Breeding

***Sonali Ambadas Aware*[1] *and Suraj Vilasrao Thakare*[2]**

[1]*Junior Research Assistant, VNMKV, Parbhani, M.S. Maharashtra, India*
[2]*District Manager, MSSCL, Akola, Maharashtra, India*

Abstract

The increasing population and its food requirement have become a burden on decreasing arable land and our food production. This situation has led plant scientists to develop products with special properties in a short time. Marker-assisted selection (MAS) has emerged as a potential tool to obtain good results in plants and increase yield in a short time with the help of molecular markers. MAS is widely used in plant breeding to perform disease identification, diversity analysis, happy superposition, gene pyramiding, various applications and cultural practices of various types of cereals, legumes, oilseeds, and fiber crops.

Keywords: *Marker, marker-assisted selection, morphological, biochemical and DNA-based*

Introduction

The main goal of plant breeding is to increase crop yield and the second goal is to improve quality, develop light- and heat-insensitive varieties, tolerance to biotic and abiotic stresses, simultaneous maturation, and water and nutrient use. Efficiency, elimination of toxins, and cultivation of different product groups allow to achieve high agricultural yields and sustainable development. Better understanding and development of genetics has increased the efficiency of plant breeding to achieve desired goals in agricultural crops. Effective and efficient use of molecular markers in crop improvement programs increases selection efficiency and productivity and accelerates the growth cycle to create new varieties with good results. Marker-assisted selection (MAS) is a technique that uses markers (morphological, biochemical, or DNA/RNA variation-based markers) to indirectly select for traits of interest (e.g., productivity, disease, genetic determinants resistance, abiotic stress tolerance and/or quality). This process is used in plant and animal breeding. The rapid development of biotechnology has allowed breeders to develop better selections to replace traditional selections based on phenotypic pedigree.

Marker-assisted selection is a type of indirect selection in which selection for a trait of interest depends not on the trait itself but on markers associated with it. For example, if the MAS is used to select people with the disease, the level of the disease is not included but the alleles associated with the disease are used to determine the presence of the disease. Linked alleles are considered to be associated with the gene of interest and/or multiple loci ([QTL] quantitative trait locus). MAS can be used for traits that are difficult to measure, have limited heritability, and/or emerge at a later stage of development.

Types of Markers

Markers can be divided into the following

- ***Morphological-based*:** These are a primary marker of things that have an effect on plant morphology. Check morphology, color, male sterility or resistance, etc. The genes affecting it have been studied in various plant species. Examples of such characters include the presence or absence of awns, leaf color, height, berry color, berry aroma, etc. can be found. Good crops such as corn, tomatoes, peas, barley, or wheat have many or hundreds of characters. These genes are assigned to different chromosomes.
- ***Biochemical-based*:** These include genes that code for proteins that can be extracted and observed; for example, isoenzymes and storage proteins.
    - Cytology: Chromosome bands produced by different dyes; for example G band
    - Biology: Different organisms or bacterial biotypes based on host–pathogen or host–parasite interactions can be used as markers, as the genetic makeup of the disease affects susceptibility to the disease or parasites.
- ***DNA and/or molecular based*:** Specific (DNA sequences) occurring near a gene or region of interest can be analyzed using various molecular techniques (RFLP [random fragment length polymorphism], RAPD [randomly amplified polymorphic DNA], AFLP [amplified fragment length polymorphism]), DAF, SCAR (sequence characterized amplified regions), microsatellites, etc.

The association of simple genetic characters with multiple traits in plants was first reported by Sax in 1923, when he observed that divergence in gene size was associated with divergence in color genes in bean (*Phaseolus vulgaris L.*). In 1935, Rasmusson found a connection between flowering time (lot) and simple background flower color in peas.

Genes versus Markers

A type of interest is directly related to the production of proteins that produce the phenotype, whereas a marker should not reflect the quality of interest but is related to genetics (if the gametes are clustered together due to a decrease in homologous recombination when separated) characters and genes of interest). In most cases, genes have been discovered and their existence can be observed directly with confidence. However, if the gene has not been isolated, the help of markers is needed to write the gene of interest. In such case there may be some false-positive results due to recombination between marker of interest and gene (or QTL). A perfect marker would elicit no false-positive results.

Important Properties of Ideal Markers for MAS

An ideal marker

- Easy recognition of all possible phenotypes (homo and heterozygotes) from all different alleles
- Demonstrates measurable differences in expression between trait types and/or gene of interest alleles, early in the development of the organism
- Has no effect on the trait of interest that varies depending on the allele at the marker loci
- Low or null interaction among the markers allowing the use of many at the same time in a segregating population
- Abundant in number
- Polymorphic

Demerits of Morphological Markers

Morphological markers are associated with several general deficits that reduce their usefulness including:

- The delay of marker expression until late into the development of the organism
- Dominance
- Deleterious effects

- Pleiotropy
- Confounding effects of genes unrelated to the gene or trait of interest but which also affect the morphological marker (epistasis)
- Rare polymorphism
- Frequent confounding effects of environmental factors which affect the morphological characteristics of the organism

To avoid problems specific to morphological markers, the DNA-based markers have been developed. They are highly polymorphic, simple inheritance (often codomimant), abundantly occur throughout the genome, easy and fast to detect, minimum pleiotropic effect, and detection is not dependent on the developmental stage of the organism. Numerous markers have been mapped to different chromosomes in several crops including rice, wheat, maize, soybean, and several others. Those markers have been used in diversity analysis, parentage detection, DNA fingerprinting, and prediction of hybrid performance. Molecular markers are useful in indirect selection, where individuals can be manually selected for subsequent breeding. Selection for major genes linked to markers

The major genes which are responsible for economically important characteristics are frequent in the plant kingdom. Such characteristics include disease resistance, male sterility, self-incompatibility; others related to shape, color, and architecture of whole plants and are often of mono- or oligogenic in nature. The marker loci which are tightly linked to major genes can be used for selection and are sometimes more efficient than direct selection for the target gene. Such vantages in efficiency may be due for example, to higher expression of the marker mRNA in such cases that the marker is actually a gene. Alternatively, in such cases that the target gene of interest differs between two alleles by a difficult-to-detect single nucleotide polymorphism, an external marker (be it another gene or a polymorphism that is easier to detect, such as a short tandem repeat) may present as the most realistic option.

Situations that are Favorable for Molecular Marker Selection

There are several indications for the use of molecular markers in the selection of a genetic trait.

In such situations

- The selected character is expressed late in plant development, like fruit and flower features or adult characters with a juvenile period (so that it is not necessary to wait for the organism to become fully developed before arrangements can be made for propagation)

- The expression of the target gene is recessive (so that individuals which are heterozygous positive for the recessive allele can be crossed to produce some homozygous offspring with the desired trait)
- There is requirement for the presence of special conditions in order to invoke expression of the target gene(s), as in the case of breeding for disease and pest resistance (where inoculation with the disease or subjection to pests would otherwise be required). This advantage is due to the fact that the error caused by the vaccination process is not possible, and many regions do not allow vaccination of the disease for safety reasons. Additionally, the problem of identifying an unstable gene environment can be avoided.
- The phenotype is influenced by two or more unlinked genes (epistasis). For example, many seeds that cause diseases or pests are also selected for the seed pyramid. While the cost of genotyping (a molecular marker test sample) is decreasing, the cost of phenotyping is increasing, especially in developing countries. As technology continues to evolve, MAS power is growing in popularity.

Marker-assisted Selection Steps Involved

In general, the first step is to first map the genes or multiple trait loci (QTL) of interest using different methods and then use this information to select service brands. In general, the markers used should be close to the gene of interest (<5 recombination units, or cM) to ensure that only a small fraction recombines in the selected population. Often, two markers rather than just one are used to reduce errors resulting from homologous recombination. For example, if two flanking markers are used together and separated by approximately 20 cM, the probability of duplicate targets is high (99%) (Fig. 1).

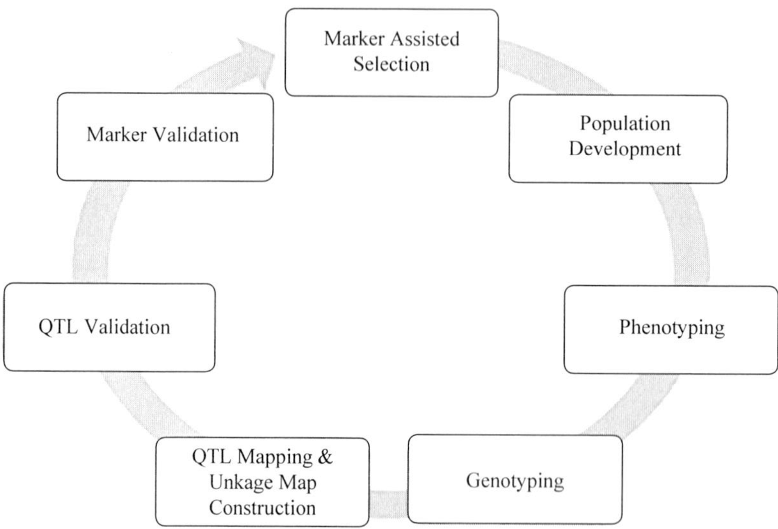

Fig. 1: Some important steps involved in marker-assisted selection. (QTL: quantitative trait locus) *Source*: Nadeem *et al.* 2018.

QTL Mapping Techniques

In plants, QTL mapping is often done using crosses; a hybrid is created from two parents with different phenotypes of interest. Commonly used recombinant inbred lines (RILs), double haploid (DH), backcross, and F2. Correlation between phenotype and markers was tested in these populations to determine the location of the QTL. This process is based on chaining and is therefore called "chain mapping".

One-Step MAS and QTL Mapping

Compared to two-step QTL mapping and MAS, one-step breeding is achieved for varieties. In this method, characters associated with the quality of interest are identified by QTL mapping during the first few breeding cycles and the same data are then used for humans. In this way, a pedigree structure is created from families formed by crossbreeding more than one parent (triple or quadruple crossbreeding). Both phenotyping and genotyping were performed using molecular markers characterizing the QTL like locus of interest. This will identify the marker and allele effect. When these advantages are identified, the response to the selection program is predicted by increasing the frequency of the alleles. Marker alleles with the desired effect can be further used in the next selection cycle or in other experiments.

High-throughput Genotyping Technology

High-throughput genotyping technology has recently been developed to enable efficient screening of multiple genotypes. This will help breeders transition from conventional breeding to selective breeding. An example of this automation is the use of DNA isolation robots, capillary electrophoresis, and pipetting robots.

A new example of a capillary system is the Applied Biosystems 3130 Genetic Analyzer. This is a new generation of 4-capillary electrophoresis measured by the laboratory as low to medium.

Backcrossing Using Marker-assisted Selection

At least five to six generations of backcrossing are required for the gene of interest to be transferred from donor to recipient (possibly unchanged by improvement (reincarnation adapts to many things). Restoration of genotypes can be accelerated by molecular markers. If F1 is heterozygous for the marker, individuals whose parents carry the marker in the first generation or in subsequent generations will still carry chromosomes marked by the marker.

Marker-assisted Gene Selection

Gene pyramids have been proposed and used to increase resistance to diseases and insects by selecting two or more genes simultaneously. For example, in rice, this pyramid was created to protect against blight and blast diseases. The advantage of using markers in this context is that markers linked to the QTL allele can be selected for the same phenotypic effect.

Conclusion

Marker-assisted selection is a technology that has already proved its value. Due to the number of QTLs, genes, and markers identified, the MAS is likely to become more valuable. It is likely to become more valuable as a larger number of genes are identified and their functions and interactions elucidated. Reduced costs and optimized strategies for integrating MAS with phenotypic selection are needed before the technology can reach its full potential.

References

Agarwal, M., Shrivastava N., Padh H. "Advances in molecular marker techniques and their applications in plant sciences." Plant Cell Reports 27,4 (2008):617-31

Ahmar, S., Gill R.A., Jung K.H., Faheem A., Qasim M.U., Mubeen M., et al. "Conventional and Molecular Techniques from Simple Breeding to Speed Breeding in Crop Plants: Recent Advances and Future Outlook." International Journal of Molecular Sciences 21, 7 (2020):2590.

Alexandratos, N., Bruinsma J. 2012. World Agriculture Towards 2030/2050: The 2012 Revision https://www.fao.org/3/ap106e/ap106e.pdf [Last accessed December, 2023].

Barcaccia, G. 2010. Molecular markers for characterizing and conserving crop plant germplasm. Molecular Techniques in Crop Improvement. Dordrecht: Springer. pp. 231-54.

Bhargava, A., Srivastava S. 2019. Toward participatory plant breeding. Participatory Plant Breeding: Concept and Applications. Singapore: Springer. pp. 69-86.

Bohar, R., Chitkineni A., Varshney R.K. "Genetic molecular markers to accelerate genetic gains in crops." BioTechniques Sep;69, 3 (September 2020):158-60.

Brennan, J.P., Rehman A., Raman H., Milgate A.W., Pleming D. 2005. Martin P.J. An economic assessment of the value of molecular markers in plant breeding programs. 49th Annual Conf. of the Australian Agricultural and Resource Economics Society. Coffs Harbour, Australia.

Brumlop, S., Finckh M.R. 2010. Applications and potentials of marker assisted selection (MAS) in plant breeding. Final report of the F+ E project "Applications and Potentials of Smart Breeding"(FKZ 350 889 0020) On behalf of the Federal Agency for Nature Conservation.

Chand, S, Chandra K, Khatik CL. 2020. Varietal Release, Notification and Denotification System in India. Plant Breeding-Current and Future Views. IntechOpen. DOI:10.5772/intechopen.94212.

Charcosset, A., Moreau L. "Use of molecular markers for the development of new cultivars and the evaluation of genetic diversity." Euphytica 137, 1 (June 2004):81-94.

Collard, B.C., Jahufer M.Z., Brouwer J.B., Pang E.C. "An Introduction to Markers, Quantitative Trait Loci (QTL) Mapping and Marker-assisted Selection for Crop Improvement: The Basic Concepts." Euphytica 142, 1-2 (2005):169-96.

Collard, B.C., Mackill D.J. "Marker-assisted Selection: An Approach for Precision Plant Breeding in the Twenty-first Century." Philosophical Transactions of the Royal Society B: Biological Sciences 363, 1491 (February 2008):557-72.

Dar, A.A., Mahajan R., Sharma S. "Molecular Markers for Characterization and Conservation of Plant Genetic Resources." The Indian Journal of Agricultural Sciences 89, 11 (November 2019):3-11.

Davey, J.W., Hohenlohe P.A., Etter P.D., Boone J.Q., Catchen J.M., Blaxter M.L. "Genome-wide Genetic Marker Discovery and Genotyping Using Next-generation Sequencing." Nature Reviews Genetics 12, 7 (2011):499-510.

Eagles, H.A., Bariana H.S., Ogbonnaya F.C., Rebetzke G.J., Hollamby G.J., Henry R.J., et al. "Implementation of Markers in Australian Wheat Breeding." Australian Journal of Agricultural Research 52, 12 (2001):1349-56.

França, L.T., Carrilho E., Kist T.B. "A Review of DNA Sequencing Techniques." Quarterly Reviews of Biophysics 35, 2 (2002):169.

Francia, E., Tacconi G., Crosatti C., Barabaschi D., Bulgarelli D., Dall'Aglio E., et al. "Marker Assisted Selection in Crop Plants." Plant Cell, Tissue and Organ Culture 82, 3 (September 2005):317-42.

Gupta, P.K., Langridge P., Mir R.R. "Marker-assisted Wheat Breeding: Present Status and Future Possibilities." Molecular Breeding 26, 2 (August 2010):145-61.

He, J., Zhao X., Laroche A., Lu Z.X., Liu H., Li Z. "Genotyping-by-sequencing (GBS), An Ultimate Marker-assisted Selection (MAS) Tool to Accelerate Plant Breeding." Frontiers in Plant Science 5 (September 2014):484.

Jiang, G.L. 2013. Molecular markers and marker-assisted breeding in plants. Plant Breeding from Laboratories to Fields. pp. 45-83. InTech. DOI:10.5772/52583.

Kage, U, Kumar A, Dhokane D, Karre S, Kushalappa AC. Functional molecular markers for crop improvement. Critical Reviews in Biotechnology. 2016 Sep 2;36(5):917-30

Karaköy, T, Baloch FS, Toklu F, Özkan H. Variation for selected morphological and quality-related traits among 178 faba bean landraces collected from Turkey. Plant Genetic Resources. 2014 Apr 1;12(1):5

Mackill, D.J., Ni J. 2001. Molecular Mapping and Marker-assisted Selection for Major-Gene Traits in Rice Rice Genetics IV. Edited by Khush G.S., Brar D.S., Hardy B. Philippines: International Rice Research Institute. pp. 137-51.

Madina, M.H., Haque M.E., Dutta A.K., Islam M.A., Deb A.C., Sikdar B. "Estimation of Genetic Diversity in Six Lentil (Lens culinaris Medik.) Varieties Using Morphological and Biochemical Markers." International Journal of Scientific & Engineering Research 4 (2013):819-25.

Mateu-Andres, I., De Paco L. "Allozymic Differentiation of the Antirrhinum majus and A. siculum Species Groups." Annals of Botany 95, 3 (February 2005):465-73.

Melchinger, A.E., Utz H.F., Schön C.C. "Quantitative Trait Locus (QTL) Mapping Using Different Testers and Independent Population Samples in Maize Reveals Low Power of QTL Detection and Large Bias in Estimates of QTL Effects." Genetics 149, 1 (May 1998):383-403.

Mir Mohammadi Maibody, S.A., Golkar P. "Application of DNA Molecular Markers in Plant Breeding." Journal of Plant Genetic Research 6, 1 (September, 2019):1-30.

Mohler, V., Singrün C. 2004. General Considerations: Marker-assisted Selection. Molecular Marker Systems in Plant Breeding and Crop Improvement Berlin: Springer. pp. 305-17.

Mondini, L., Noorani A., Pagnotta M.A. "Assessing Plant Genetic Diversity by Molecular Tools." Diversity 1, 1 (2009):19-35.

Mullis, K., Faloona F., Scharf S., Saiki R.K., Horn G.T., Erlich H. "Specific Enzymatic Amplification of DNA In Vitro: The Polymerase Chain Reaction." Cold Spring Harbor Symposia on Quantitative Biology 51 (January 1986):263-73.

Nadeem, M.A., Nawaz M.A., Shahid M.Q., Doğan Y., Comertpay G., Yıldız M., et al. "DNA Molecular Markers in Plant Breeding: Current Status and Recent Advancements in Genomic Selection and Genome Editing." Biotechnology & Biotechnological Equipment 32, 2 (March 2018):261-85.

Nakaya, A., Isobe S.N. "Will Genomic Selection Be a Practical Method for Plant Breeding?" Annals of Botany. 110, 6 (November 2012):1303-16.

Platten, J.D., Cobb J.N., Zantua R.E. "Criteria for Evaluating Molecular Markers: Comprehensive Quality Metrics to Improve Marker-assisted Selection." PloS One 14, 1 (January 2019): e0210529.

Ribaut, J.M., Hoisington D. "Marker-assisted Selection: New Tools and Strategies." Trends in Plant Science 3, 6 (1998):236-9.

Semagn, K., Bjørnstad Å., Ndjiondjop M.N. "An Overview of Molecular Marker Methods for Plants." African Journal of Biotechnology 5, 25 (2006).

Sharp, P.J., Johnston S., Brown G., McIntosh R.A., Pallotta M., Carter M., et al. "Validation of Molecular Markers for Wheat Breeding." Australian Journal of Agricultural Research 52, 12 (2001):1357-66.

Singh, A.K., Gopalakrishnan S., Singh V.P., Prabhu K.V., Mohapatra T., Singh N.K., et al. "Marker Assisted Selection: A Paradigm Shift in Basmati Breeding." Indian Journal of Genetics and Plant Breeding 71, 2 (May 2011):120.

Singh, B.D., Singh A.K. 2015. Marker-assisted Plant Breeding: Principles and Practices. New Delhi: Springer.

Tabor, H.K., Risch N.J., Myers R.M. "Candidate-Gene Approaches for Studying Complex Genetic Traits: Practical Considerations." Nature Reviews Genetics 3, 5 (2002):391-7.

Bailey, D.C. 1983. Isozymic Variation and Plant Breeders' Rights. Isoenzymes in Plant Genetics and Breeding, part A. Edited by Tanksley S.D., Orton T. Amsterdam: Elsevier. pp. 425-41.

Tester, M., Langridge P. "Breeding Technologies to Increase Crop Production in a Changing World." Science 327, 5967 (February 2010):818-22.

Uzun, A.Y., Yesiloglu T., Aka-Kacar Y., Tuzcu O., Gulsen O. "Genetic Diversity and Relationships within Citrus and Related Genera Based on Sequence-related Amplified Polymorphism Markers (SRAPs)." Scientia Horticulturae 121,3 (July 2009):306-12.

Vos, P., Hogers R., Bleeker M., Reijans M., Lee T.V., Hornes M., et al. "AFLP: A New Technique for DNA Fingerprinting. Nucleic Acids Research. 23, 21 (1995):4407-14.

Watson, A., Hickey L.T., Christopher J., Rutkoski J., Poland J., Hayes B.J. "Multivariate Genomic Selection and Potential of Rapid Indirect Selection with Speed Breeding in Spring Wheat." Crop Science59, 5 (September 2019):1945-59.

Weiland, J.J., Yu M.H. "A Cleaved Amplified Polymorphic Sequence (CAPS) Marker Associated with Root-knot Nematode Resistance in Sugarbeet." Crop Science 43, 5 (September 2003):1814-8.

Wenzl, P., Carling J., Kudrna D., Jaccoud D., Huttner E., Kleinhofs A., et al. "Diversity Arrays Technology (DArT) for Whole-genome Profiling of Barley." Proceedings of the National Academy of Sciences101, 26 (June 2004):9915-20.

Williams, J.G., Kubelik A.R., Livak K.J., Rafalski J.A., Tingey S.V. "DNA Polymorphisms Amplified by Arbitrary Primers are Useful as Genetic Markers." Nucleic Acids Research 18, 22 (November 1990):6531-5.

Wu, J., Zhang M., Zhang X., Guo L., Qi T., Wang H., et al. "Development of InDel Markers for the Restorer Gene Rf1 and Assessment of their Utility for Marker-assisted Selection in Cotton." Euphytica213,11 (November 2017):251.

Xu, Y., Zhang X.Q., Harasymow S., Westcott S., Zhang W., Li C. "Molecular Marker-assisted Backcrossing Breeding: An Example to Transfer a Thermostable β-Amylase Gene from Wild Barley." Molecular Breeding 38,5 (2018):63.

Yadawad A., Gadpale A., Hanchinal R.R., Nadaf H.L., Desai S.A., Biradar S., et al. "Pyramiding of Leaf Rust Resistance Genes in Bread Wheat Variety DWR 162 through Marker Assisted Backcrossing." Indian Journal of Genetics and Plant Breeding 77 (2017):251-7.

Yang L., Fu S., Khan M.A., Zeng W., Fu J. "Molecular Cloning and Development of RAPD-SCAR Markers for Dimocarpus Longan Variety Authentication." SpringerPlus 2, 1(2013):501.

Zane, L., Bargelloni L., Patarnello T. "Strategies for Microsatellite Isolation: A Review." Molecular Ecology 11, 1 (2002):1-6.

Zhou, W.C., Kolb F.L., Bai G.H., Domier L.L., Boze L.K., Smith N.J. "Validation of a Major QTL for Scab Resistance with SSR Markers and Use of Marker-assisted Selection in Wheat." Plant Breeding 122, 1 (2003):40-6.

9

Genome Editing: Approaches Principles and Application in Plant Breeding

Sajal Saha[1], Deepa Bhadana[2], Rinkey Arya[3] and Rajib Das[4]

[1]Ph.D. Nagaland University, School of Agricultural Sciences, Medziphema, Nagaland India

[2]Ph.D. Chaudhary Charan Singh University, Meerut, Uttar Pradesh, India

[3]Faculty of Agriculture and Agroforestry, DSB Campus Nainital, Kumaun University Uttarakhand, India

[4]Assistant Professor, College of Agriculture, Central Agricultural University Pasighat Arunachal Pradesh, India

Abstract

Genome editing has revolutionized plant breeding, offering precise and efficient methods to enhance crop traits. This chapter provides an overview of genome editing's principles, applications, challenges, and future prospects in plant breeding. The chapter explores genome editing technologies, with a focus on CRISPR-Cas9 (clustered regularly interspaced short palindromic repeats-associated protein 9), TALENs (transcription activator-like effector nucleases), and ZFNs (zinc finger nucleases). CRISPR-Cas9, in particular, has become the go-to tool for precision breeding, allowing scientists to target and modify specific genes with unprecedented accuracy. Applications of genome editing in plant breeding are diverse. It enables the development of crops with enhanced disease resistance, drought tolerance, and improved nutritional content. Additionally, genome-edited crops reduce environmental impact by requiring fewer chemical inputs and promoting sustainable farming practices. Accelerated breeding shortens the time needed to create new crop varieties, addressing urgent agricultural challenges. Looking ahead, the future of genome editing holds exciting possibilities. Continuous advancements in technology promise even more precise editing tools and high-throughput capabilities. Precision breeding will become the norm, allowing for tailored crop varieties and customized agricultural practices. Despite challenges, genome editing offers hope for increased food security, sustainability, and resilience in agriculture, ultimately benefiting a growing global population while preserving our planet's resources.

Keywords: *Genome editing, CRISPR-Cas9, genes, transgenic breeding*

Introduction

The field of plant breeding plays a pivotal role in agriculture, aiming to enhance crop traits for increased yield, nutritional value, and resistance to pests and diseases. Over centuries, traditional breeding methods have been used to develop new crop varieties by selectively crossing plants with desired traits. However, these methods can be time-consuming and imprecise. Throughout history, humans have engaged in agricultural practices and have consciously or unconsciously selected and bred plants to improve their characteristics. The concept of plant breeding dates back to the early agricultural civilizations, where farmers saved and replanted seeds from plants with desirable traits. Notable figures such as Gregor Mendel and his work on pea plants laid the foundation for understanding inheritance patterns in the late 19th century. The mid-20th century saw the advent of modern breeding techniques, including hybridization and mutagenesis, which significantly accelerated the development of new crop varieties. However, these methods still had limitations in terms of precision. In recent years, a revolutionary breakthrough has occurred in the form of genome editing technologies. Genome editing allows for the precise modification of specific genes within an organism's DNA. One of the most prominent and versatile techniques is the CRISPR-Cas9 (clustered regularly interspaced short palindromic repeats-associated protein 9) system, which was adapted from a natural defense mechanism in bacteria. This technology enables scientists to target and edit specific genes in plant genomes with unprecedented accuracy and efficiency. The emergence of genome editing has the potential to revolutionize plant breeding by providing breeders with a powerful tool to directly and precisely manipulate the genetic makeup of crops. This chapter delves into the principles and applications of genome editing in the context of plant breeding, exploring the transformative possibilities it offers for improving crop traits, increasing sustainability, and addressing global food security challenges.

Genome Editing Technologies

CRISPR-Cas9: Components and Mechanism

The CRISPR-Cas9 system is a groundbreaking genome editing technology that has transformed the field of genetics and biotechnology. In 2020, Emmanuelle Charpentier and Jennifer Doudna were jointly awarded the Nobel Prize in chemistry for the development of the CRISPR-Cas9 gene-editing technology. It consists of two main components:

- ***CRISPR:*** These are specific DNA sequences found in bacteria that have been derived from viral DNA. They serve as a memory system for the bacterium to recognize and defend against viruses.
- ***Cas9 or CRISPR-associated protein 9:*** Cas9 is an enzyme that acts as molecular scissors. It can be programmed to target and cut specific DNA sequences in a highly precise manner.

The mechanism of CRISPR-Cas9 involves the following steps:

1. Designing a guide RNA (gRNA) molecule that is complementary to the target DNA sequenceCas9 is loaded with the gRNA, which guides it to the target DNA.
2. Cas9 then creates a double-stranded break at the target DNA site.
3. The cell's repair machinery can introduce changes (edits) during the repair process (Fig. 1).

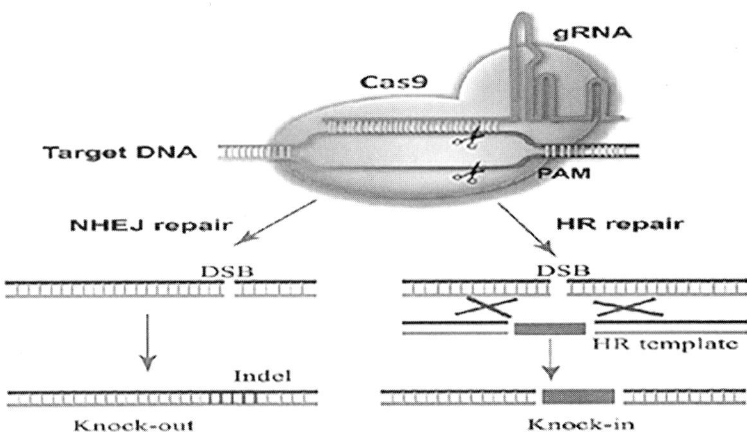

Fig. 1: Working system of CRISPR-Cas9 (clustered regularly interspaced short palindromic repeats-associated protein 9). (DSB: double-strand break; gRNA: guide RNA; HR: homologous recombination; NHEJ: nonhomologous end joining; PAM: ??)

Source: Tavakoli *et al.* 2021.

Other Genome Editing Techniques: TALENs and ZFNs

While CRISPR-Cas9 is the most prominent genome editing technology, it is essential to be aware of alternative methods. Two such methods are transcription activator-like effector nucleases (TALENs) and zinc finger nucleases (ZFNs):

- ***TALENs:*** These are custom-designed proteins that can be engineered to target specific DNA sequences. They consist of a DNA-binding domain derived from transcription activator-like effectors (TALEs) and a nuclease domain that induces DNA breaks at the target site. TALENs were widely used before the emergence of CRISPR-Cas9 but are less commonly employed now due to the latter's simplicity.

- ***ZFNs:*** These are another class of engineered proteins used for genome editing. They consist of zinc finger domains, which bind to specific DNA sequences, and a nuclease domain that creates breaks in the DNA. Like TALENs, ZFNs were widely used before CRISPR-Cas9 but are less favored today due to the ease and precision of CRISPR technology. While CRISPR-Cas9 dominates the field of genome editing, TALENs and ZFNs remain viable options for specific applications or when designing gRNAs for CRISPR is challenging. Understanding the principles and capabilities of these technologies is crucial for plant breeders and geneticists (Fig. 2).

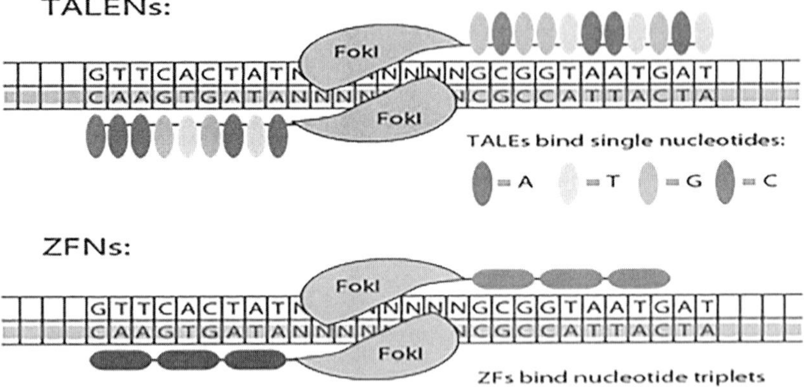

Fig. 2: Working system of transcription activator-like effector (TALE) nucleases (TALENs) and zinc finger nucleases (ZFNs).

Source: Xenbase. New genome editing technologies. [online] Available from https://www.xenbase.org/xenbase/static-xenbase/CRISPR.jsp [Last accessed December, 2023]

- ***CyDENT:*** In our book chapter, we introduced CyDENT base editing, which is an innovative approach that stood apart from traditional TALE-based tools and CRISPR-based methods for editing nuclear and organellar DNA. Unlike some other techniques that relied on double-stranded DNA deaminases, which could edit substrate bases on both DNA strands and thereby reduced editing precision, CyDENT offered a CRISPR-free, strand-selective, and modular base editing solution. CyDENT's core components included a pair of TALEs fused with a FokI nickase, a cytidine deaminase that specifically targeted single strands of DNA, and an exonuclease that collaboratively generated a single-stranded DNA substrate ready for deamination. This unique setup allowed for precise base editing in nuclear, mitochondrial, and chloroplast genomes. The experiments demonstrated the effectiveness of CyDENT base editing across these genomes. Particularly, at specific mitochondrial sites, we achieved impressive editing efficiencies of up to 14% while maintaining a remarkable strand specificity of 95%. Furthermore, by swapping out the CyDENT deaminase with one that had a preference for editing Guanine and Cytosine bases motifs, we successfully accomplished up to 20% mitochondrial base editing at sites that were previously inaccessible using other editing methods (Hu *et al.* 2023).

Achievements

- ***Increase yield:*** In a range of climatic conditions, editing the C terminus of *Oryza sativa* LOGL5, which encodes a cytokinin activation enzyme in rice, increased crop productivity (Wang *et al.* 2020). CRISPR-Cas-mediated editing of additional genes, such as O. sativa PIN5b (controlling panicle size), *O. sativa* GS3 (regulating grain size) (Zeng *et al.* 2019), *Triticum aestivum* GW2 (Zhang *et al.* 2018), *O. sativa* GW2 (Zhou *et al.* 2019), and O. sativa GW5 (regulating grain weight) (Liu *et al.* 2017), has also resulted in higher-yielding crop plants (Khatodia *et al.*, 2016).
- ***Improved quality:*** Low-gluten wheat lines with up to 85% lower immune reactivity have been developed using CRISPR-Cas to target the conserved area of gluten genes (Sánchez-León *et al.* 2018). CRISPR-Cas has also made it easier to develop high-quality crops with higher levels of carotenoid (Dong *et al.* 2020) and aminobutyric acid (Li *et al.* 2018), lower levels of phytic acid (Khan *et al.* 2019), and higher levels of oleic acid (Do *et al.* 2019).
- ***Disease-resistant:*** Rice lines with broad-spectrum resistance to *Xanthomonas oryzae* pv. *oryzae* were developed using CRISPR-Cas9 to modify the promoter region of *O. sativa* SWEET11, *O. sativa* SWEET13, and O. sativa SWEET14 (Xu *et al.* 2019; Oliva *et al.* 2019). CRISPR-Cas mutation of three wheat homologues of EDR1 resulted in plants with better resistance to *Blumeria graminis* f. sp. *tritici* (Zhang *et al.* 2017). CRISPR-Cas to target *Solanum lycopersicum* mildew resistance locus O 1 (MLO1) conferred tolerance to powdery mildew-causing oidium neolycopersici in tomato (Nekrasov *et al.* 2017), and simultaneous mutation of all three MLO homologues resulted in a wheat variety with broad-spectrum powdery mildew resistance (Wang *et al.* 2014).

Principles of Genome Editing in Plants

Cas9 Delivery Methods

The Cas9 protein and gRNA need to be introduced into plant cells to initiate the genome editing process. Two commonly used delivery methods are:

- ***Agrobacterium-mediated transformation:*** *Agrobacterium tumefaciens* is a bacterium used to transfer genes into plant cells. In this method, scientists engineer *Agrobacterium* to carry the Cas9 and gRNA. When this modified *Agrobacterium* infects plant cells, it transfers the genome editing components, leading to targeted DNA modification. This method is particularly popular for editing the genomes of dicot plants (Fig. 3).

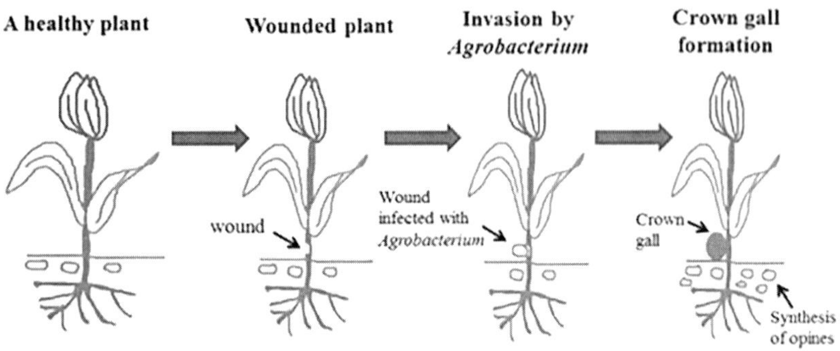

Fig. 3: Tumor induced by *Agrobacterium* in plants.

Source: ICAR eCourse. Principles of Plant Biotechnology. https://agrimoon.com/wp-content/uploads/Principles-of-Plant-Biotechnology.pdf [Last accessed December, 2023].

- ***Molecular basis of Agrobacterium-mediated transformation*:** The virulent strains of *A. tumefaciens* harbor large plasmids (140–235 kbp) known as tumor-inducing (Ti) plasmid involving elements like TDNA, virulence (Vir) region, origin of replication, region enabling conjugative transfer, and o-cat region (required for catabolism of opines).
- ***TDNA:*** It is a small, specific segment of the plasmid, about 24 kb in size, and found integrated in the plant nuclear DNA at random site. This DNA segment is flanked by right and left borders.
- ***Genes on TDNA:***
    - **The TDNA contains two groups of genes, which possess the ability to express in plants as:** Oncogenes for synthesis of auxins and cytokinins (phytohormones). The overproduction of phytohormones leads to proliferation of callus or tumor formation.

- Opine synthesizing genes for thc synthesis of opines (a product from amino acids and sugars secreted by the crown gall infected cells and utilized by *A. tumefaciens* as carbon and nitrogen sources). Thus, opines act as source of nutrient for bacterial growth, e.g., octopine and nopaline.

- ***Transfer of TDNA and integration into the plant cell:***
  - The transfer of TDNA to the plant cell is mediated by type-IV secretion system composed of proteins encoded by VirB and VirD4 that form a conjugative pilus (T-pilus). VirD4 serves as a "linker" that helps in the interaction of the processed T-DNA/VirD2 complex with the VirB-encoded pilus. Other Vir genes (VirE2, VirE3, VirF, VirD5) also pass through this T-pilus to aid in the assembly of TDNA/Vir protein complex in the plant cytoplasm forming a mature T-complex.
  - Most VirB proteins help in the formation of the membrane channel or act as adenosine triphosphatases (ATPases) to provide energy for assembly and export processes of channel. VirB proteins, including VirB2, VirB5, and VirB7 help in the formation of the T-pilus. VirB2 is the major pilin protein that undergoes processing and cyclization. Single-stranded TDNA (Ss-TDNA) is coated with VirE2, a nonsequence specific single-stranded DNA-binding protein.
  - VirD2 and VirE2 protect the ss-T strand from nucleases inside the plant cytoplasm by attaching to the 5'end. Both VirD2 and VirE2 proteins have nuclear localization signals (NLS) which serves as pilot proteins to guide the mature Tcomplex to the plant nucleus.
  - The efficiency of transfer is enhanced by VirC2 proteins, which recognize and bind to the overdrive enhancer element.
  - Some additional proteins like importins, VIP1, and VirF may interact with the Tstrand, either directly or indirectly, to form larger T-complexes in the plant cell. VirF directs the proteins coating T-complex (VIP1 and VirE2) for destruction in proteasome.
  - Inside the nucleus, ss-TDNA is converted into double-stranded TDNA (ds-TDNA) which gets integrated into the plant genome via process called illegitimate recombination (Fig. 4).

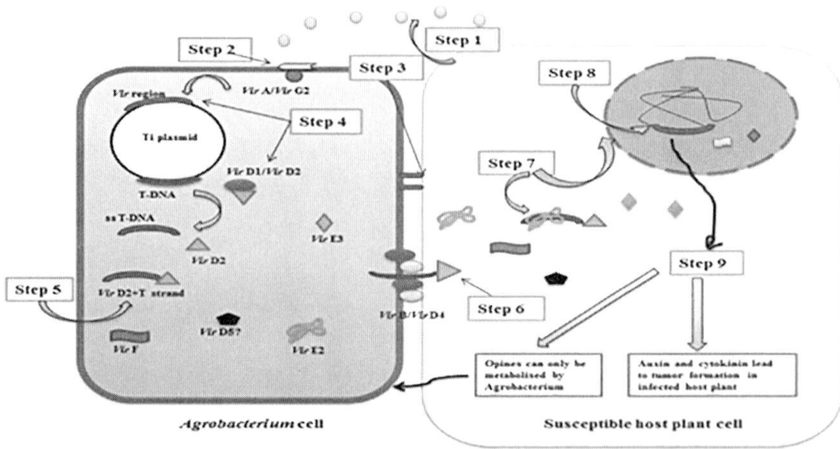

Fig. 4: Schematic representation of *Agrobacterium*-mediated transformation. Step 1: production of signal molecules form wounded plant cell; step 2: recognition of signal molecules by bacterial receptors; step 3: attachment of *Agrobacterium* to plant cell; step 4: activation of virulence (Vir) proteins which process single-stranded TDNA (ss-TDNA); step 5: formation of immature T-complex; step 6: TDNA transfer; step 7: assembly of mature T-complex and nuclear transport; step 8: random T-DNA integration in the plant genome; step 9: expression of bacterial genes and synthesis of bacterial proteins.

Source: Adapted and modified from Pacurar *et al.* 2011.

Biolistic Bombardment

Prof. Sanford and colleagues at Cornell University (USA) developed the original bombardment concept in 1987 and coined the term "biolistics" (short for "biological ballistics") for both the process and the device. This method, often referred to as the "gene gun" approach, involves coating microscopic gold or tungsten particles with Cas9 and gRNA and then shooting them into plant cells using a biolistic device. The particles penetrate plant cell walls and deliver the editing components into the cell's nucleus. Biolistic bombardment is effective for editing a wide range of plant species, including monocots and dicots.

Procedure of Particle Bombardment Gun

The biolistic gun employs the principle of conservation of momentum and uses the passage of helium gas through the cylinder with arrange of velocities required for optimal transformation of various cell types. It consists of a bombardment chamber which is connected to an outlet for vacuum creation. The bombardment chamber consists of a plastic rupture disk below which macrocarrier is loaded with microcarriers. These microcarriers consist of gold or tungsten micro pellets coated with DNA for transformation (Fig. 5).

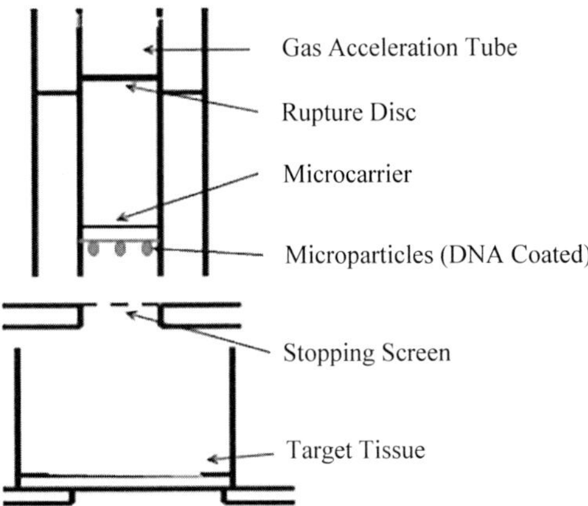

Fig. 5: Procedure of particle bombardment gun.

Source: ICAR eCourse. Principles of Plant Biotechnology. https://agrimoon.com/wp-content/uploads/Principles-of-Plant-Biotechnology.pdf [Last accessed December, 2023].

Target Selection

Selecting the right target genes is a crucial step in genome editing. Consider the following aspects:

- ***Criteria for target genes:*** Choose genes that play a pivotal role in the desired trait or characteristic. These genes should have a known function and a clear link to the trait you aim to modify. Additionally, consider the specificity of the target gene to minimize off-target effects.
- ***Off-target effects:*** Genome editing can sometimes unintentionally affect DNA sequences other than the intended target. It is essential to use bioinformatics tools and thorough screening to predict and minimize off-target effects. Choose gRNAs with minimal off-target potential to ensure precise editing.

Repair Mechanisms

After a double-strand DNA break is induced by Cas9, plant cells employ repair mechanisms to mend the damage. Two primary repair mechanisms are involved:

1. ***Non-homologous end joining (NHEJ)*:** NHEJ is an error-prone repair mechanism that directly joins the broken DNA ends together. This process often results in insertions or deletions (indels) at the repair site. Indels can disrupt the target gene's function, leading to gene knockout or modification. NHEJ is the default repair pathway in most plant cells.
2. ***Homology-directed repair (HDR)*:** HDR is a precise repair mechanism that relies on a template DNA molecule with a sequence matching the

target gene. Scientists can use HDR to introduce specific changes or insertions at the target site by providing a template DNA along with Cas9 and gRNA. HDR is less common in plant cells than NHEJ, but is valuable for precise gene editing and gene insertion. Understanding these principles is essential for successful genome editing in plants, as it guides the choice of delivery methods, target genes, and repair strategies for specific breeding goals.

Applications in Plant Breeding

Crop Improvement

- ***Disease resistance:*** Genome editing enables the development of crops with enhanced resistance to pathogens and pests. By targeting specific genes involved in plant defense mechanisms, breeders can create crops that are more resilient to diseases, reducing the need for chemical pesticides.
- ***Drought tolerance:*** Drought is a significant threat to global agriculture. Genome editing can be used to modify plants to better withstand water scarcity by targeting genes involved in water-use efficiency, root architecture, and stress responses.
- ***Nutritional enhancement:*** Genome editing can be applied to improve the nutritional content of crops. For example, scientists can enhance the levels of essential vitamins and minerals in staple crops, addressing malnutrition and dietary deficiencies in vulnerable populations.

Reducing Environmental Impact

- ***Sustainable agriculture:*** By developing crops with increased disease resistance and drought tolerance, farmers can achieve higher yields with fewer inputs, such as pesticides and water. This promotes sustainable agricultural practices and reduces the environmental footprint of farming.
- ***Reduced chemical usage:*** Genome-edited crops with built-in resistance to pests and diseases require fewer chemical interventions. This not only reduces the use of synthetic chemicals, but also decreases the risk of chemical residues in food and the environment.

Accelerated Breeding

Genome editing accelerates the traditional breeding process in several ways:

- ***Shortened breeding cycles:*** Traditional breeding can take many years to develop new crop varieties. Genome editing allows for the rapid creation of desired traits, significantly shortening breeding cycles and getting improved crops to market more quickly.

- ***Rapid trait introduction:*** Instead of waiting for favorable traits to naturally occur or be bred into a crop, genome editing enables breeders to introduce specific traits quickly. This precision breeding approach allows for the immediate incorporation of desired traits into existing varieties, saving time and resources. These applications demonstrate the transformative potential of genome editing in plant breeding, offering solutions to some of the most pressing challenges in agriculture, including food security, sustainability, and resilience in the face of environmental changes.

Challenges and Ethical Considerations

Regulatory Framework

The development and deployment of genome-edited crops are subject to a complex regulatory landscape that varies from country to country. Two critical aspects of this regulatory framework are:

- ***Approval and commercialization:*** Genome-edited crops must undergo rigorous safety assessments and regulatory approval processes before they can be commercialized. Regulatory agencies evaluate factors such as the environmental impact, food safety, and potential allergenicity of these crops. The approval process can be time-consuming and costly, creating challenges for researchers and breeders eager to bring new varieties to market.
- ***Labeling and traceability:*** Once genome-edited crops are approved for commercialization, there is often a requirement for clear labeling and traceability. This means that consumers have the right to know if a product contains genome-edited ingredients. Establishing effective labeling and traceability systems can be logistically challenging and requires coordination across the supply chain. The regulatory landscape continues to evolve, and it's essential for stakeholders in the genome editing and agriculture industries to navigate these regulations effectively to ensure the safe and responsible development and commercialization of genome-edited crops. Understanding and addressing these regulatory challenges is crucial to the successful adoption of genome-edited crops and the responsible management of their introduction into the market. It involves a delicate balance between promoting innovation and ensuring safety and transparency for consumers and the environment.

Ethical and Societal Considerations

Social Acceptance

- ***Public perception and trust:*** Genome editing technologies can be met with varying degrees of acceptance among the public. Some concerns may include perceived "genetic tampering", potential long-term health effects, or ethical objections related to manipulating the DNA of organisms. Public education and engagement are crucial to build trust and understanding about genome editing.
- ***Equity and access:*** Ensuring that the benefits of genome-edited crops are equitably distributed is an ethical consideration. It is important to address issues of access for small-scale farmers and communities in developing countries to avoid exacerbating existing inequalities in agriculture.
- ***Transparency and inclusivity:*** Ethical practices in genome editing include transparent communication about the technology, potential risks, and benefits. Inclusivity in decision-making processes ensures that diverse perspectives are considered, fostering responsible innovation.

Ecological Implications

The ecological impact of genome-edited crops is another area of concern:

- ***Gene flow:*** Genome-edited crops may potentially crossbreed with wild or nonedited varieties, leading to unintended ecological consequences. Monitoring and managing gene flow are important to prevent unintended environmental impacts.
- ***Biodiversity:*** Altering the genetics of crops can affect the biodiversity of ecosystems. Genome editors must consider how their modifications might impact local flora and fauna, including pollinators and other wildlife.
- ***Unintended consequences:*** There is always the potential for unintended ecological consequences when introducing genetically modified organisms. Risk assessments and environmental impact studies are essential to minimize these risks. Addressing these ethical and societal considerations involves a multidisciplinary approach that includes scientists, policymakers, ethicists, and the public. Engaging in transparent and informed discussions, assessing risks, and implementing responsible practices are essential steps to ensure that genome-edited crops are developed and deployed in a manner that aligns with societal values and ecological sustainability.

Future Directions

Advancements in Genome Editing

- ***New techniques and tools:*** Researchers are continually developing new genome editing techniques and tools that offer increased precision, efficiency, and versatility. These may include innovations in Cas proteins, novel delivery methods, and improved gRNA design. For example, base editing and prime editing are emerging technologies that allow for even more precise changes to DNA sequences.
- ***High-throughput editing:*** High-throughput genome editing platforms are becoming increasingly sophisticated. These platforms enable the simultaneous editing of multiple genes or the screening of a vast number of genetic variations. High-throughput approaches are revolutionizing the pace at which scientists can study gene function and discover new traits for crop improvement. The future of genome editing in plant breeding will likely see a convergence of these advancements, making it easier to create tailored crop varieties with multiple desired traits quickly and efficiently. As genome editing technologies become more refined and accessible, their applications in agriculture are expected to expand, contributing to increased food security, sustainability, and resilience in the face of global challenges. Researchers and breeders should stay informed about these advancements to harness their full potential for crop improvement.

Precision Breeding

Precision breeding represents a paradigm shift in plant breeding, which offers the ability to create highly tailored crop varieties and customize agriculture practices:

- ***Tailored crop varieties:*** Genome editing allows for the precise modification of specific genes, resulting in crop varieties designed to meet specific needs. Whether it is improving disease resistance, nutritional content, or stress tolerance, precision breeding enables the creation of crops optimized for particular environments or markets.
- ***Customized agriculture:*** Precision breeding goes hand in hand with customized agriculture practices. Farmers can select or develop crop varieties that are ideally suited to their local conditions, optimizing resource use, and reducing environmental impacts. Precision agriculture technologies, such as remote sensing and data analytics, further enhance the ability to customize farming practices.

Conclusion

Genome editing has revolutionized plant breeding by providing precise and efficient tools to modify the genetic makeup of crops. This chapter has explored the principles and applications of genome editing in plant breeding, highlighting its potential to address global food security challenges, reduce environmental impact, and accelerate the breeding process. Looking ahead, the future of genome editing in plant breeding is promising. Advances in technology, including new techniques and high-throughput editing, will continue to expand the possibilities for crop improvement. Precision breeding will become increasingly prevalent, leading to the development of highly specialized crop varieties tailored to diverse agricultural needs. However, ethical, regulatory, and societal considerations will play a crucial role in shaping the direction of genome editing in agriculture. Engaging in transparent and inclusive discussions and addressing concerns about safety and equity will be essential for responsible innovation in this field. As genome editing technologies mature and our understanding of plant genetics deepens, we can anticipate a future where agriculture becomes more efficient, sustainable, and resilient, ultimately contributing to the global challenge of feeding a growing population while preserving our planet's resources.

References

Do, P.T., Nguyen C.X., Bui H.T., Trans L.T.N., Stacey G., Gillman J.D., et al. "Demonstration of Highly Efficient Dual gRNA CRISPR/Cas9 Editing of the Homeologous GmFAD2-1A and GmFAD2-1B Genes to Yield a High Oleic, Low Linoleic and Alpha-linolenic Acid Phenotype in Soybean." BMC Plant Biology 19 (2019):311.

Dong, O.X., Yu S., Jain R., Zhang N., Duong P.Q., Butler C., et al. "Marker-free Carotenoid-Enriched Rice Generated through Targeted Gene Insertion Using CRISPR-Cas9." Nature Communications 11(2020):1178.

Hu, J., Sun Y., Li B., Liu Z., Wang Z., Gao Q., et al. "Strand-preferred Base Editing of Organellar and Nuclear Genomes Using CyDENT." Nature Biotechnology (2023):1-10.

Khan, M.S.S., Basnet R., Islam S.A. Shu, Q. "Mutational Analysis of OsPLDalpha1 Reveals its Involvement in Phytic Acid Biosynthesis in Rice Grains." Journal of Agricultural and Food Chemistry 67 (2019):11436-43.

Khatodia, S., Bhatotia K., Passricha N., Khurana S.M.P., Tuteja N. "The CRISPR/Cas Genome-editing Tool: Application in Improvement of Crops." Frontiers in Plant Science 7 (2016):506.

Li, R., Li R., Li X., Fu D., Zhu B., Tian H., et al. "Multiplexed CRISPR/Cas9-mediated Metabolic Engineering of Gamma-aminobutyric Acid Levels in Solanum Lycopersicum." Plant Biotechnology Journal. 16 (2018):415-27.

Liu, J., Chen J., Zheng X., Wu F., Lin Q., Heng Y., et al. "GW5 Acts in the Brassinosteroid Signalling Pathway to Regulate Grain Width and Weight in Rice." Nature Plants 3 (2017):17043.

ICAR eCourse. Principles of Plant Biotechnology. https://agrimoon.com/wp-content/uploads/Principles-of-Plant-Biotechnology.pdf [Last accessed December, 2023].

Nekrasov, V., Wang C., Win J., Lanz C., Weigel D., Kamoun S. "Rapid Generation of a Transgene Free Powdery Mildew Resistant Tomato by Genome Deletion." Scientific Reports 7 (2017):482.

Oliva, R., Ji C., Atienza-Grande G., Huguet-Tapia J.C., Perez-Quintero A., Li T., et al. "Broad-spectrum Resistance to Bacterial Blight in Rice Using Genome Editing." Nature Biotechnology 37 (2019):1344-50.

Pacurar, D.I., Thordal-Christensen H., Pacurar M.L., Pamfil D., Botez C., Bellini C. "Agrobacterium tumefaciens: From Crown Gall Tumors to Genetic Transformation." Physiological and Molecular Plant Pathology 76, 2 (2011):76-81.

Sánchez-León, S., Gil-Humanes J., Ozuna C.V., Giménez M.J., Sousa C., Voytas D.F., et al. "Low-gluten, Nontransgenic Wheat Engineered with CRISPR/Cas9." Plant Biotechnology Journal. 16 (2018):902-10.

Tavakoli, K., Pour-Aboughadareh A., Kianersi F., Poczai P., Etminan A., Shooshtari L. "Applications of CRISPR-Cas9 as an Advanced Genome Editing System in Life Sciences." BioTech 10, 3 (2021): 14.

Wang, Y., Cheng X., Shan Q., Zhang Y., Liu J., Gao C., et al. "Simultaneous Editing of Three Homoeoalleles in Hexaploid Bread Wheat Confers Heritable Resistance to Powdery Mildew." Nature Biotechnology 32 (2014):947-51.

Xenbase. New genome editing technologies. [online] Available from https://www.xenbase.org/xenbase/static-xenbase/CRISPR.jsp [Last accessed December, 2023].

Xu, Z., Xu X., Gong Q., Li Z., Li Y., Wang S., et al. "Engineering Broad-spectrum Bacterial Blight Resistance by Simultaneously Disrupting Variable TALE-binding Elements of Multiple Susceptibility Genes in Rice." Molecular Plant 12 (2019):1434-46.

Zeng, Y., Wen J., Zhao W., Wang Q., Huang, W. "Rational Improvement of Rice Yield and Cold Tolerance by Editing the Three Genes OsPIN5b, GS3, and OsMYB30 with the CRISPR-Cas9 system." Frontiers Plant Science 10 (2019):1663.

Zhang, Y., Bai Y., Wu G., Zou S., Chen Y., Gao C., et al. Simultaneous modification of three homoeologs of TaEDR1 by genome editing enhances powdery mildew resistance in wheat. The Plant Journal: For Cell and Molecular Biology 91 (2017):714-24.

Zhang, Y., Li D., Zhang D., Zhao X., Cao X., Dong L., et al. "Analysis of the Functions of TaGW2 Homoeologs in Wheat Grain Weight and Protein Content Traits." The Plant Journal: For Cell and Molecular Biology 94 (2018):857-66.

Zhou, J., Xin X., He Y., Chen H., Li Q., Tang X., et al. "Multiplex QTL Editing of Grain Related Genes Improves Yield in Elite Rice Varieties. Plant Cell Reports 38 (2019):475-85.

10

Advance Genomic Technologies TILLING and Eco-TILLING

Sivabharathi R. C.[1], Anandhi K.[2], Muthuswamy A.[3] and Raiza Christina G.[4]

[1,4]Ph.D. Scholar, Centre for Plant Breeding and Genetics, Tamil Nadu Agricultural University, Coimbatore, Tamil Nadu, India

[2]Assistant Professor, Department of Pulses, Centre for Plant Breeding and Genetics, Tamil Nadu Agricultural University, Coimbatore, Tamil Nadu, India

[3]Associate Professor, Agricultural College and Research Institute, Tamil Nadu Agricultural University, Karur, Tamil Nadu, India

Abstract

The whole genome sequencing helps the researcher to identify the desired gene and make it useful for the breeding program by transferring that particular gene into improved cultivar. Since, the whole genome sequence is not available for all crops and animals; there is a need of advanced techniques to overcome this problem. One such technique is TILLING (Targeting Induced Local Lesions IN Genomes) and its related technique called Eco-TILLING. TILLING is a reverse genetic approach used to identify the point mutation created by chemical mutagen in the mutagenized population. The chemical mutagen is most widely used and it will create allelic series of mutations which include nonsense, missense, or splice site mutation. The individuals with mutation were pooled by two-dimensional or three-dimensional pooling techniques. These mutations were identified by the formation of heteroduplex complex in the mutated region of the gene. Then, the heteroduplex is cleaved by S1 nuclease family enzyme and were examined in gel electrophoresis which distinguishes heteroduplex from homoduplexes. Now, the identified mutants in gel electrophoresis were subjected to sequencing for confirmation. Eco-TILLING is similar to the TILLING except for the mutant population, as we use the natural mutant population in Eco-TILLING; whereas in TILLING we use the artificial mutant population and also pooling is done with reference sample in Eco-TILLING. The advantage of TILLING is that it is a cost-effective, rapid screening of mutant population and a nontransgenic approach as it does not require biosafety measures.

The main disadvantages of these techniques are low rate of mutation induction and the need of skillful labor.

Keywords: *Eco-TILLING, EMS, heteroduplex, Li-COR, TILLING*

Introduction

The new era of genomics began in 1977 with introduction of Sanger (chain termination) and Maxam and Gilbert (chemical) sequencing methods and it extended up to the sixth-generation sequencing techniques (Kchouk *et al.* 2017). One of the most direct ways of establishing gene function is to identify a mutation in a specific gene and to link this mutation to the phenotypic change in the mutated organism. Forward and reverse genetics are the two different approaches in functional genomics to identify the gene sequence and its function. Forward genetics uses mutant phenotype with random mutation to identify the gene sequence and it includes map-based cloning, association mapping, genome-wide association studies, etc. Reverse genetic approach use the gene or protein sequence with known or unknown function to determine the function of gene by analyzing the phenotype and approaches used are insertional mutagenesis, RNA interference, CRISPR/Cas system (Kumar *et al.* 2013), zinc finger nuclease, transcription activator-like effector nucleases (TALENs), TILLING (Targeting Induced Local Lesions IN Genomes), and Eco-TILLING (Aklilu 2021).

TILLING is a reverse genetic approach that combines chemical mutagenesis with polymerase chain reaction (PCR)-based screening to identify the point mutations in the region of interest (McCallum 2000). TILLING was first begun in the late 1990s from the effort of a graduate student, Claire McCallum, (and collaborators from Fred Hutchinson Cancer Research Center and Howard Hughes Medical Institute), who worked on characterizing the function of two chromomethylase genes (CMT genes) in *Arabidopsis*. McCallum utilized reverse genetic approaches such as transfer DNA (T-DNA) lines and antisense RNA, but was unable to characterize CMT2. Later, he worked on the principle of the TILLING and was successful with TILLING approach.

Why TILLING is Preferred Over the Reverse Genetic Approaches (Tadele 2016)

- Both natural and mutagenized population can be screened. However, mutagenized population has a large number of randomly distributed mutations per genome.
- Both missense and nonsense mutations can be detected.
- Plant heterozygous for a mutation can be detected.
- There is no need of transgenic manipulation.

- There are no biosafety problems.
- It is cost-effective than genetic engineering.
- It bypasses sophisticated tissue culture barriers.

Principle and Steps Involved in TILLING Approach

Principle: Gene segments are amplified using gene-specific primers and the products are denatured and reannealed to form heteroduplex between the mutated sequence and its wild-type counterparts. These heteroduplexes are subjected to cleavage by the CEL I endonuclease. The cleaved products are analyzed on denaturating polyacrylamide gel electrophoresis (PAGE) (Khan *et al.* 2018).

Steps Involved in TILLING (Khan *et al.* 2018)

1. Development of mutagenized population
2. DNA preparation and pooling of individuals
3. *Mutation discovery*:
    a. PCR amplification of gene of interest which are fluorescently tagged primers
    b. Denaturation and annealing forms heteroduplex at site of mutation
    c. Detection of heteroduplex by CEL I endonuclease, which cleaves at heteroduplex mismatches
    d. Identification of the mutant individual—cleaved products is detected on PAGE (Li-COR analyzer).
    e. Sequencing of mutant PCR products

Development of Mapping Population

The ideal mutagen for TILLING is one that randomly induces single nucleotide substitutions, or small insertions/deletions (~<30 nucleotides) at a high frequency in the genome. The chemical mutagen ethyl methane sulfonate (EMS) generates mostly single nucleotide polymorphisms (SNPs) and can be controlled to produce a high density of point mutations, causing a variety of lesions including nonsense and missense mutations.

The effect of treatment with EMS is highly predictable; G:C > A:T transition changes represent the majority of induced mutations in most organisms (Muqaddasi and Arif 2012). This is especially striking in *Arabidopsis and wheat* where >99% of mutations identified by TILLING is G:C > A:T transitions (Weil *et al.* 2005).

The density of induced mutations is a major factor in the efficiency and cost of mutation discovery. To discover 10 mutations in Maize TILLING populations, with a density of 1 mutation per 500 kb, will take approximately twice as long and cost twice as much as to discover the same number of mutations as in *Arabidopsis* TILLING populations with a density of ~1 mutation per 250 kb (Weil *et al.* 2005). Differences in mutation density between organisms may result from differences in the uptake of cytotoxic response and repair of lesions induced by treatment with the mutagen. Following established mutagenesis protocols would therefore seem a good route to success, however, an approximately 1.5-fold range in mutation density was observed when following a standard protocol for *Arabidopsis seed mutagenesis* (Weil *et al.* 2005).

Seed and Pollen Mutagenesis

Seed and pollen mutagenesis is demonstrated in Figure 1.

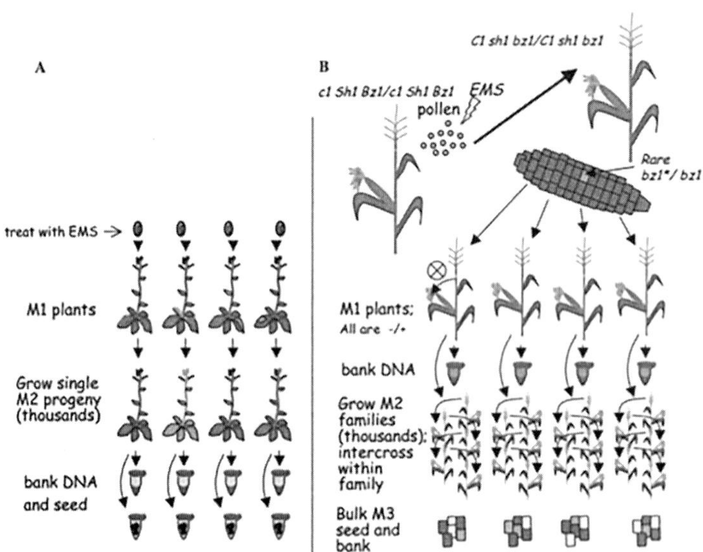

Fig. 1: Mutagenesis scheme. *A*, Seed mutagenesis; *B*, pollen mutagenesis. (EMS: ethyl methane sulfonate)

Source: Weil *et al.* 2005.

DNA Preparation and Pooling of Individuals

Sample pooling will directly affect the efficiency and cost of mutation discovery. The screening of four samples pooled together will take approximately twice as long and cost twice as much as screening a pool of eight samples. Factors that affect the ability to pool include the quality of genomic DNA, the accuracy of sample quantification, and the method used for SNP discovery. Various TILLING groups have performed screens utilizing two-, three-, four-, six-, and eightfold pooling.

Two basic pooling strategies have been most often used (Till, Zerr, *et al.* 2006):

1. One-dimensional pooling strategy where each individual sample is represented in only one pool; when a mutation is identified in a pool of eight individuals, each member of the pool is then screened independently to identify the individual harboring the mutation.
2. Two-dimensional pooling where each sample is present in two unique pools; although duplicating each sample reduces the throughput of detection in pools by half, the sample harboring the mutation is unambiguously determined in the pool screening step, so that there is no need to screen individual samples as it is done in the one-dimensional strategy. Two-dimensional pooling involves screening each sample with twofold coverage and potential false-positive and false-negative errors are minimized at the initial screening step, rather than when individuals are screened in the second step with one-dimensional pooling.

Mutation Discovery

The SNP discovery technologies include array-based methods, Sanger sequencing denaturing high-performance liquid chromatography (HPLC), mass spectroscopy, denaturing gradient capillary electrophoresis, and enzymatic mismatch cleavage. The most common method used for TILLING has been enzymatic mismatch cleavage and resolution on polyacrylamide gels to detect the cleaved fragments.

- ***PCR amplification* (Till, Colbert, *et al.* 2006):** The objective of TILLING is to identify plants with mutations in the target gene, and most mutations in noncoding sequences such as promoters, introns, and untranslated regions will have no effect on gene function. So, the target sequence should be chosen to minimize such sequences and coding sequences.

  As a first step, PCR forward and reverse primers are designed to amplify 1,500 base pairs or less of genomic DNA from a locus of interest. The forward and reverse primers are differentially 5′-end labeled with IRD700 and IRD800 dye labels for fluorescent detection at approximately 700 nm and 800 nm, respectively.

- ***Heteroduplex formation***: The mutated region created by either insertion or deletions (in TILLLING it is related to insertions) may lead to the formation of heteroduplex. The wild and mutant amplified DNA fragments are mixed and they are subjected to denaturation at 95°C and followed by reannealing at room temperature (White *et al.* 1992).
- ***Detection of heteroduplexes* (Till, Zerr *et al.* 2006):** CEL I nuclease is the most commonly used nuclease even though a number of nucleases have been described for mismatch cleavage (Oleykowski *et al.* 1998). In addition to CEL I nuclease, mung bean nuclease is also recommended for TILLING (Till, Zerr, *et al.* 2006). Celery (*Apium graveolens*) is a common vegetable used for extraction of CEL I enzyme. Heteroduplex formed after reannealing of the PCR products were treated with CEL I enzyme. Single-stranded specific endonuclease which cleaves 3′ site of single base mismatch in the heteroduplexes (Fig. 2).

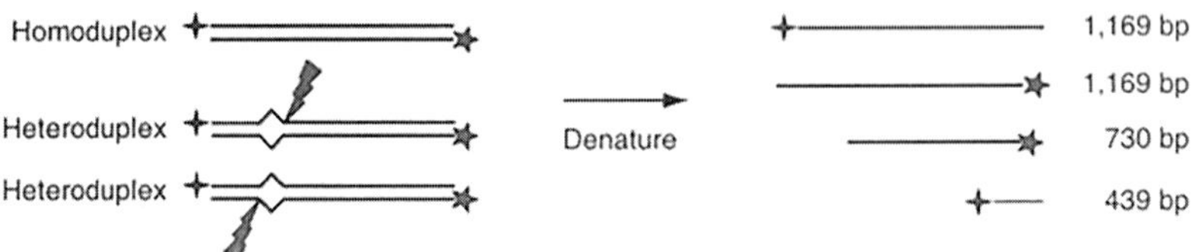

Fig. 2: Schematic diagram of mismatch cleavage.
Source: Till, Zerr *et al.* 2006

- ***Identification of mutant individual***: CEL I-treated PCR products are resolved on PAGE and screened for the presence of novel fragments. The amplified wild and mutant products that are denaturated will run on the gel producing different banding pattern. The wild type is not broken as it is not mutated, so it has high density and will move slowly on the gel; whereas the mutated DNA are broken by CEL I nuclease, have less density compared to wild and so will move faster.

The result will be the wild type in the starting point of gel with single band whereas the mutated DNA observed at the end of gel with two bands. In Li-COR analyzer, as primers are fluorescently labeled with IRD700 and IRD800 the results will be in two separate gels which makes easier identification (Fig. 3).

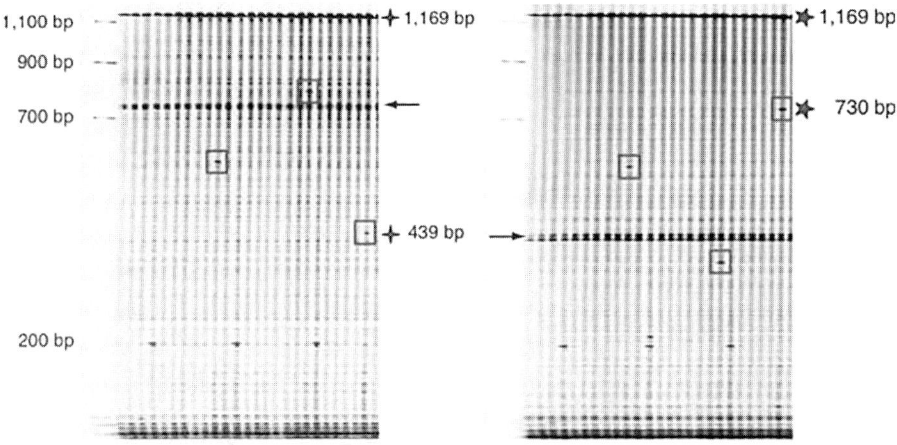

Fig. 3: IRdye 700 (*left*) and IRdye 800 (*right*) in which true mutations are marked in blue boxes and red boxes respectively.

Source: Till, Zerr, *et al.* 2006.

- ***Sequencing of mutant PCR products***: The mutants identified by the gel electrophoresis are allowed for sequencing of those particular PCR products for confirmation with the gel results. The overall general procedure of tilling is described in Figure 4.

Applications of TILLING (De-Kai *et al.* 2006)

- ***Functional genomics***: TILLING helps to study the functions of all specific gene sequences, their expression and location in organism. Construction of TILLING library helps scientists to search for mutations in gene of interest without having first knowledge of gene product.
- ***Genetic engineering***: TILLING produces phenotypic variants without introducing foreign DNA of any type into a plant's genome. T-DNA/ transposon insertions are used to obtain specific gene knockouts but practically limited to some crops only.

  TILLING is in front of transgene, as consists of identification of numerous mutations within a targeted region of whole genome.
- ***Genetic diversity***: Alternative to wild relatives—TILLING is used to introduce useful genetic variation of elite germplasm and it is also applicable in a population which has several preexisting polymorphism for developing SNPs.

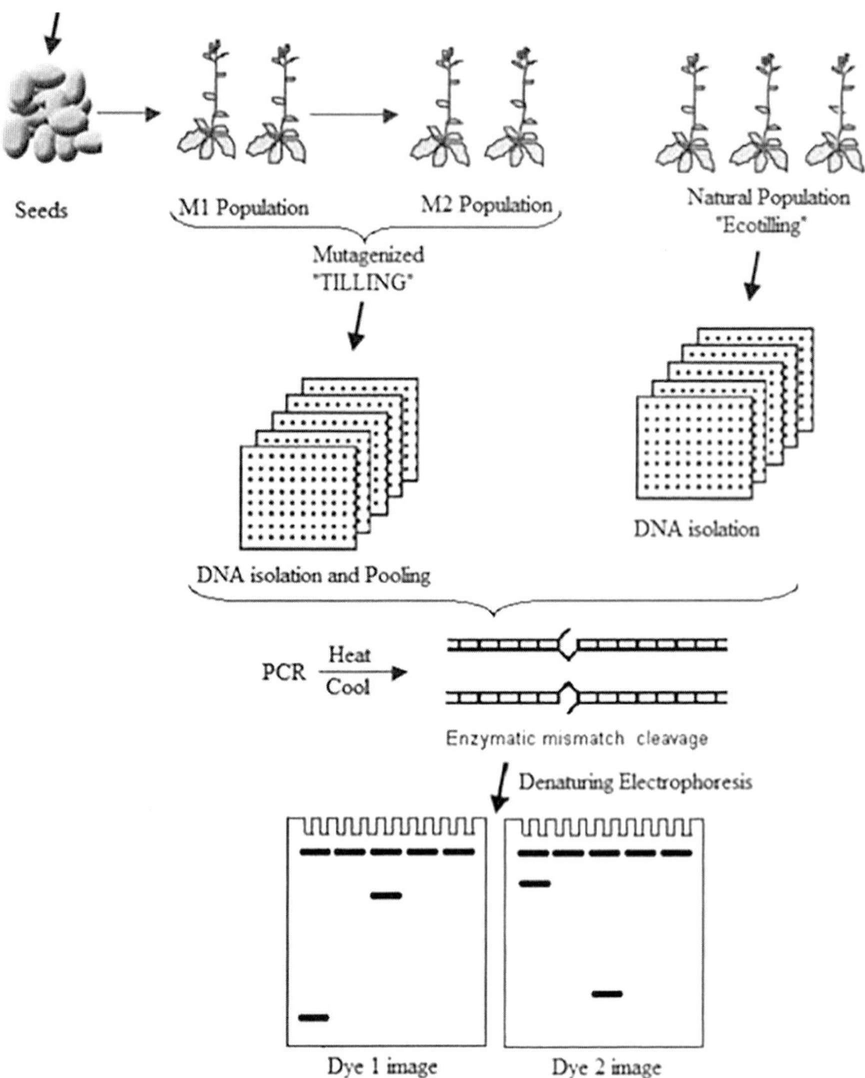

Fig. 4: Basic steps for typical TILLING and Eco-TILLING.(PCR: polymerase chain reaction; TILLING: Targeting Induced Local Lesions IN Genomes)

Source: Simsek and Kaçar 2010

Merits of TILLING

- TILLING is independent of genome size, reproductive system, or generation time.
- High throughput and data analysis can also be automated.

- Valuable for essential genes, where sublethal alleles are required for phenotypic analysis.
- TILLING is suitable for any organism that can be heavily mutagenized, even those that lack genetic tools.
- For organisms that do not have efficient transformation systems, TILLING is the only practical choice.
- It overcomes disadvantage of knockout gene therapy for a specific gene, as it eliminates the need for removal of gene from gene pool to detect its function.
- TILLING can introduce genetic variation in an elite germplasm without need to acquire variation and thus avoiding introduction of agriculturally undesirable traits.
- The likelihood of recovering a deleterious mutation can be calculated in advance.
- It is highly sensitive and cost-effective.
- It detects SNP over thousands of samples.
- It is used in producing phenotypic variants without introducing foreign DNA of any type into a plant's genome.
- It has the advantage of exemption from regulatory approval requirements which is strictly obliged for transgenic crops when commercially suitable variations are discovered.
- It involves rapid and low-cost discovery of induced point mutations in populations of chemically mutagenized individuals from diverse organisms.
- There is no requirement of transgenic or sophisticated tissue culture methodology.
- It is applicable to any organism.

Demerits of TILLING

- TILLING requires skillful labor.
- It depends on primer designing.
- Mutagenized organism must be kept alive long enough to screen mutant population in vegetatively propagated plant species.
- Starting with a homozygous population is desirable.
- Rate of mutation induction is low.

Eco-TILLING

The enzymatic mismatch cleavage method used for TILLING is applicable to any heteroduplexed DNA target regardless of the source of the nucleotide polymorphism. Therefore, the same methods are applicable to the discovery of natural nucleotide variation in populations.

The modification of TILLING for the discovery and genotyping of natural nucleotide polymorphisms was termed Eco-TILLING (Cooper *et al.* 2013). The first publication of the Eco-TILLING method in which TILLING was modified to mine for natural polymorphisms was in 2004 from work in *Arabidopsis thaliana* (Till, Zerr, *et al.* 2006).

- Eco-TILLING is similar to TILLING, except that its objective is to identify natural genetic variation as opposed to induced mutations.
- Many species are not amenable to chemical mutagenesis, therefore, Eco-TILLING can aid in the discovery of natural variants and their putative gene function.
- The pooling is done with the reference sample in case of Eco-TILLING.

This approach allows one to rapidly screen many samples with gene of interests to identify naturally occurring SNPs and/or small insertions/deletions (Ins/DELs) (Fig. 5).

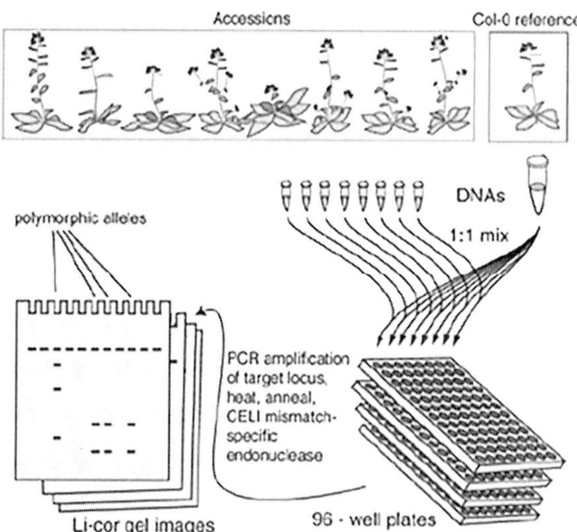

Fig. 5: Schematic representation of Eco-TILLING process. (TILLING: Targeting Induced Local Lesions IN Genomes)

Source: Comai *et al.* 2004.

Applications of Eco-TILLING (Barkley and Wang 2008)

- Eco-TILLING is a fast, low-cost, and an efficient method for the discovery of nucleotide polymorphism.
- It can be used for tagging and classifying accessions based on nucleotide polymorphism.
- It can be used for functional gene analysis.
- It can be used for discovery of rare haplotypes.
- New markers will be generated from Eco-TILLING projects.
- Nonsynonymous SNPs may be identified that provide a beneficial phenotype.
- For species in which mutagenesis is impractical, exploiting natural nucleotide diversity will be invaluable for crop improvement.
- Many species are not amenable to chemical mutagenesis; Eco-TILLING can aid in the discovery of natural variants and their putative gene function.
- Following the Eco-TILLING screen, allele-specific polymorphism can be used as marker to screen a segregating population generated from a cross with the accession carrying a promising haplotype.
- Eco-TILLING reduces the time and effort for SNP discovery by weeding out identical haplotypes.
- This method requires sequence of all individuals in a population to identify polymorphisms, which can be a burdensome expense and time-consuming.
- It also has the advantage of detecting multiple polymorphisms in a single fragment because CEL I will digest only a small proportion of the heteroduplex at a single position.

Conclusion

The techniques of TILLING and Eco-TILLING are very much useful than the genetic engineering as they do not need to follow any biosafety regulations and will not produce transgenic crop. They can be used for any crop species irrespective of their pollination behavior as well as help overcome the problems of tissue culture. In addition these techniques help identify the SNP that provide a beneficial phenotype.

References

Aklilu, E. "Review on Forward and Reverse Genetics in Plant Breeding." All Life 14, 1 (2021):127-35.

Barkley, N.A., Wang M.L. "Application of TILLING and EcoTILLING as Reverse Genetic Approaches to Elucidate the Function of Genes in Plants and Animals." Current Genomics 9, 4 (June 2008):212-26.

Comai, L., Young K., Till B.J., Reynolds S.H., Greene E.A., Codomo C.A., et al. "Efficient Discovery of DNA Polymorphisms in Natural Populations by Ecotilling." The Plant Journal 37, 5 (March 2004):778-6.

Cooper, J.L., Henikoff S., Comai L., Till B.J. 2013. "TILLING and Ecotilling for Rice." Methods in Molecular Biology 956 (2013):39-56.

Kchouk, M., Gibrat J.F., Elloumi M. "Generations of Sequencing Technologies: from First to Next Generation." Biology and Medicine 9, 3 (January 2017).

Khan, A., Abidi I., Bhat M.A., Dar Z., Ali G., Shikari A.B., et al. 2018. "Tilling and Eco-tilling–A Reverse Genetic Approach for Crop Improvement." International Journal of Current Microbiology and Applied Science 7, 06 (2018):15-21.

Kumar, A.P.K., Boualem A., Bhattacharya A., Parikh S., Desai N., Zambelli A., et al. "SMART–Sunflower Mutant population And Reverse genetic Tool for crop improvement." BMC Plant Biology 13 (2013):1-8.

McCallum, C.M., Comai L., Greene E.A., Henikoff S. "Targeted Screening for Induced Mutations." Nature Biotechnology 18, 4 (2000):455-7.

Muqaddasi, Q.H., Arif M. "Ethyle Methane Sulphonate (EMS) Induced Mutagenic Attempts to Create Genetic Variability in Basmati Rice." Journal of Plant Breeding and Crop Science 4, 7 (2012):101-5.

Oleykowski, C.A., Mullins C.R.B., Godwin AK., Yeung A.T. "Mutation detection using a novel plant endonuclease." Nucleic Acids Research 26, 20 (1998):4597-602.

Simsek, Ö., Kaçar Y.A. "Discovery of Mutations with TILLING and ECOTILLING in Plant Genomes." Scientific Research and Essays 5, 24 (2010):3799-802.

Tadele, Z. "Mutagenesis and TILLING to Dissect Gene Function in Plants." Current Genomics 17, 6 (2016):499-508.

Till, B.J., Colbert T., Codomo C.A., Enns L.C., Johnson J., Reynolds S.H., et al. "High-throughput TILLING for Arabidopsis." Arabidopsis Protocols (2006):127-35.

Till, BJ., Zerr T., Comai L., Henikoff S. "A Protocol for TILLING and Ecotilling in Plants and Animals." Nature Protocols 1, 5 (2006):2465-77.

Wang D.K., Sun Z.X., Tao Y.Z. "Application of TILLING in Plant Improvement." Acta Genetica Sinica 33, 11 (2006):957-64.

Weil, C., Monde R.A., Till B.J., Comai L., Henikoff S. "Mutagenesis and Functional Genomics in Maize." Maydica 50, 3 (2005):415-24.

White, M.B., Carvalho M., Derse D., O'Brien S.J., Dean M. "Detecting Single Base Substitutions as Heteroduplex Polymorphisms." Genomics 12, 2 (1992):301-6.

11

Advancements in Forward and Reverse Genetics

Nikki Kumari[1] , Pooja Kanwar Shekhawat[2], Dharm Veer Singh[3] and Kamaluddin[4]

[1]*Division of Genetics, Indian Agricultural Research Institute, New Delhi, India*

[2]*Department of Plant Breeding and Genetics, Sri Karan Narendra Agriculture University, Jobner, Rajasthan, India*

[3,4]*Department of Genetics and Plant Breeding, Banda University of Agriculture and Technology, Banda, Uttar Pradesh, India*

Abstract

Two cornerstone methodologies in genetics, forward genetics and reverse genetics, illuminate the intricate connection between gene sequences and their functions. Forward genetics traditionally entailed laborious steps, beginning with the observation of mutants and culminating in the identification of causal genetic alterations. While effective, this approach was time-consuming and limited in scalability. Recent innovations have revolutionized forward genetics with next-generation sequencing (NGS), genome-wide association studies (GWAS), CRISPR (clustered regularly interspaced short palindromic repeats)-based screens, and advanced computational techniques. These advancements deliver precision and efficiency, enabling researchers to explore gene function on an unprecedented scale. In contrast, reverse genetics initiates its journey armed with sequence data and harnesses the manipulation of genes to observe resultant phenotypic shifts. Traditional reverse genetics, though invaluable, faced challenges like limited precision and scalability. Cutting-edge approaches have transformed the landscape. Techniques such as prime editing, base editing, high-throughput functional genomics, single-cell genomics, and synthetic biology offer unparalleled precision and capacity. They empower scientists to scrutinize gene function at the single-cell level, conducting precise genome editing and large-scale functional genomics studies. These strides in forward and reverse genetics have charted new territories for genetic research. They hold the potential to decode intricate genetic mechanisms, propel medical advancements, bolster crop enhancement, and deepen our comprehension of evolution. Combining ancestral wisdom with contemporary technologies, genetics continues its odyssey into the genetic code, promising to unveil the hidden mysteries of life encoded within.

Keywords: *Clustered regularly interspaced short palindromic repeats (CRISPR), genome-wide association studies, next-generation sequencing, single-cell genomics*

Introduction

The world of genetics has journeyed from its humble beginnings in the garden of an Austrian monk, Gregor Mendel, to a modern landscape characterized by breathtaking discoveries and technological marvels. Mendel's meticulous pea plant experiments hinted at the existence of genes, laying the first stepping stones on a path that would ultimately lead us to deciphering the genetic code of life itself. However, the story of genetics is not one of a single leap but a series of transformative advancements that have illuminated the complexities of the genetic world. These advances have expanded the toolkit for genetic research and revolutionized fields such as medicine, agriculture, and evolutionary biology. The advent of genome sequencing, complemented by computational and empirical annotation techniques, has facilitated the identification and characterization of genes within various plant species (Schneeberger *et al.*, 2014). These developments have paved the way for the elucidation of how these genes collectively govern the intricacies of plant organisms. In the realm of genetics, two fundamental methodologies serve as the keystones for establishing the intricate relationship between a gene sequence and its functional role: forward genetics and reverse genetics. To investigate gene activities, researchers predominantly employ two strategies: forward genetics and reverse genetics. These complementary approaches have catalyzed the molecular analysis of known and previously undiscovered genes that modulate diverse plant traits.

Forward Genetics and Reverse Genetics

Forward genetics is a research approach that centers its inquiry on the identification of sequence alterations responsible for a particular mutant phenotype (Peters *et al.*, 2003). This investigative path is initiated with the observation of a naturally occurring mutant or the deliberate creation of a phenotypic variant through mutagenesis techniques (phenotype to gene approach). The pivotal objective of forward genetics is the subsequent identification of the specific sequence modifications within the genome that are causally linked to the observed phenotype. The process often entails mapping and sequencing the genetic loci associated with the trait of interest. As such, forward genetics constitutes a valuable tool for unraveling the genetic basis of distinct phenotypic traits.

Conversely, reverse genetics embarks on the journey of genetic exploration with a different starting point, sequence information retrieved from genome and transcript profiling endeavors. This approach leverages the wealth of genomic data to gain insights into gene function. It involves the strategic manipulation of specific genes to observe the resultant phenotypic consequences. Researchers may employ techniques such as gene silencing, gene knockout, or gene editing to intervene directly with the gene's activity (Raingam *et al.*, 2018). By inducing these controlled genetic alterations, scientists can meticulously decipher the gene's function within the intricate biological context. Thus, reverse genetics serves as a powerful means to bridge the gap between sequence information and functional understanding, enabling the elucidation of gene roles with precision.

Traditional Forward Genetics Techniques

The traditional technique for studying forward genetics primarily involved a series of steps, including:

- ***Phenotypic observation*:** Researchers would begin by carefully observing natural or induced mutants within a population of organisms (e.g., plants or animals). These mutants exhibited distinctive phenotypic traits or characteristics of interest.
- ***Mutagenesis*:** In some cases, researchers induced mutations using physical or chemical agents, such as radiation or chemicals like ethyl methane sulfonate (EMS). This mutagenesis step was employed to deliberately create a population of mutants with genetic diversity and novel traits.
- ***Phenotypic screening*:** Following mutagenesis, large populations of mutants were screened for specific phenotypic changes or variations. Researchers meticulously observed and documented any noticeable differences from the wild-type or normal organisms.
- ***Linking phenotype to genotype*:** After identifying mutants with intriguing phenotypes, researchers conducted genetic mapping and sequencing. Traditional mapping techniques, such as linkage analysis, and sequencing methods, often Sanger sequencing, were employed to pinpoint the genomic regions associated with these traits. This step aimed to identify the genetic alterations (mutations) responsible for the observed phenotypes.
- ***Functional analysis*:** Once the causative mutations were identified, further studies were conducted to investigate the function of the affected genes or genomic regions. This could involve gene expression analysis,

protein function studies, or complementation tests, which aimed to elucidate the specific roles of the genes in question.

- ***Confirmation and validation***: Findings were validated through additional experiments, such as genetic crosses or complementation tests. These experiments served to confirm the relationship between the identified mutations and the observed phenotypes, providing robust evidence for the genetic basis of the traits.
- ***Gene function elucidation***: Ultimately, the goal was to understand the role and function of the genes underlying the mutant phenotypes. This understanding shed light on their contributions to specific biological processes.

The traditional technique for forward genetics, while valuable, had several limitations. These are described as follows:

- ***Time-consuming and labor-intensive***: Researchers often had to manually screen and phenotype large populations of mutants, which could be a painstaking and slow procedure.
- ***Limited scalability***: Analyzing a large number of mutants was challenging, and it was difficult to systematically study multiple genes simultaneously.
- ***Dependency on natural mutations***: Relied on naturally occurring mutations or induced mutations through random mutagenesis. This reliance on serendipitous mutations made it challenging to study specific genes of interest systematically.
- ***Difficulty in identifying causative mutations***: It required extensive mapping and sequencing efforts.
- ***Lack of precision***: Researchers had limited control over the mutations generated, which could complicate the study of gene function.
- ***Limited accessibility to mutants***: Researchers often had to generate their own mutant populations, which could be resource-intensive.
- ***Ethical considerations***: Use of mutagenic agents raised ethical concerns, especially in organisms with long generation times or in ecological studies.
- ***Inefficiency in non-model organisms***: The technique was less efficient in non-model organisms, where genetic tools and resources were limited.
- ***Environmental variability***: Phenotypic variations in mutants could sometimes be influenced by environmental factors, making it challenging to distinguish genetic effects from environmental effects.

Recent Forward Genetics Techniques

Recent techniques used for forward genetics have significantly improved the efficiency, precision, and scalability of gene discovery compared to previous methods. Some of these techniques and their advantages over traditional approaches include:

- ***Next-generation sequencing (NGS)*:** NGS technologies have transformed forward genetics by enabling whole-genome sequencing (WGS) and whole-exome sequencing (WES). We can use these methods to identify genetic variants associated with a trait or disease.
- ***Genome-wide association studies (GWAS)*:** GWAS involves scanning the genomes of large cohorts of individuals to identify genetic variants, typically single-nucleotide polymorphisms (SNPs), associated with specific traits or diseases. This approach has led to numerous discoveries of genetic links to various traits and diseases.
- ***Phenome-wide association studies (PheWAS)*:** PheWAS is similar to GWAS but explores the associations between genetic variants and a wide range of phenotypes, allowing researchers to uncover unexpected connections between genetic variants and different health conditions.
- ***Clustered regularly interspaced short palindromic repeats (CRISPR)-based functional genomics screens*:** CRISPR/Cas9 technology can be used for forward genetics by systematically knocking out or activating genes across the genome. Researchers can then screen for phenotypic changes to identify genes associated with specific traits.
- ***Functional genomics databases*:** Comprehensive databases of gene function, such as CRISPR knockout libraries or RNA interference (RNAi) libraries, enable researchers to systematically target and study individual genes to discover their roles in specific traits or biological processes.
- ***Machine learning and artificial intelligence (AI)*:** Advanced computational methods, including machine learning and AI, are increasingly used to analyze large-scale genomic data in forward genetics studies. These techniques help identify candidate genes and pathways associated with phenotypes.
- ***Linkage analysis*:** While not as high-throughput as GWAS, linkage analysis remains a valuable approach in forward genetics. It involves studying genetic markers in families to identify regions of the genome associated with inherited traits.

- ***Functional genomics tools***: Various functional genomics tools, such as RNA-Seq (RNA sequencing) and proteomics, are used to assess changes in gene expression and protein levels associated with specific traits.
- ***CRISPR epigenome editing***: CRISPR-based epigenome editing allows researchers to modify epigenetic marks, such as DNA methylation or histone modifications, at specific genomic loci to study their effects on gene expression and phenotypes (Flowchart 1).

Flowchart 1: The stepwise processes involved in both *A*, traditional forward genetics technique and *B*, recent forward genetics techniques showcasing the transition from conventional methods to more advanced and data-driven approaches. (CRISPR-Cas9: clustered regularly interspaced short palindromic repeats–associated protein 9; EMS: ethyl methane sulfonate)

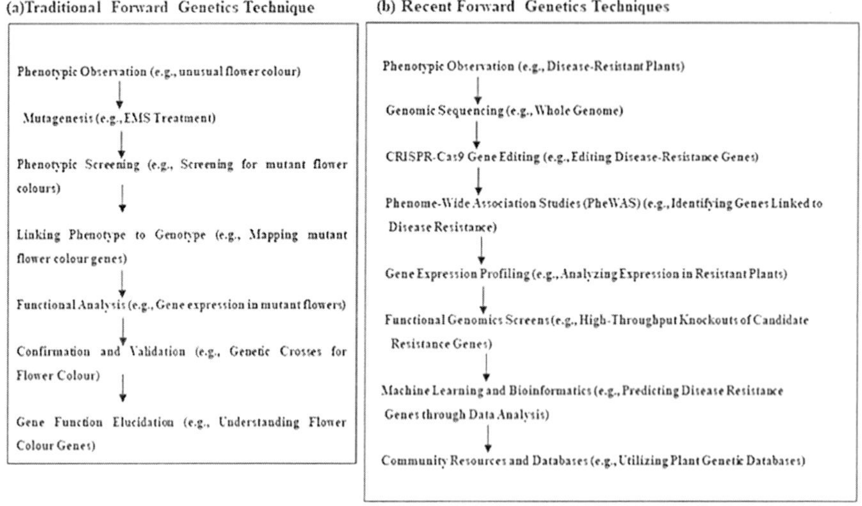

Traditional Reverse Genetics Techniques

The traditional techniques for studying reverse genetics involve manipulating an organism's genetic material to understand the function of specific genes. These techniques are essential for uncovering the role of genes in various biological processes. Here are some traditional methods for studying reverse genetics:

- ***Gene knockout***: This technique involves disrupting or eliminating the function of a specific gene within an organism. This can be achieved through various methods, including homologous recombination in mice (gene targeting) or using RNAi in other organisms. By observing the effects of the missing or nonfunctional gene, researchers can infer its role in development, physiology, or disease.

- ***Gene overexpression***: In contrast to knockout studies, overexpression involves increasing the expression of a specific gene. This is often done by introducing extra copies of the gene into an organism or by using strong promoters to drive its expression. Overexpression studies help researchers understand the function of a gene by observing the consequences of increased gene activity.
- ***Site-directed mutagenesis***: Researchers can create specific mutations in a gene of interest using techniques like polymerase chain reaction (PCR)-based mutagenesis or CRISPR/Cas9 genome editing. By introducing targeted mutations, scientists can investigate the effects of specific genetic changes on an organism's phenotype.
- ***RNAi***: This is a technique used to selectively silence the expression of specific genes by introducing small RNA molecules that inhibit the translation or stability of the gene's mRNA. This method is particularly useful for studying gene function in organisms where traditional knockout techniques are challenging or time-consuming.
- ***Transposon insertion mutagenesis***: Transposons are DNA sequences that can move around the genome, potentially disrupting the function of genes they insert into. Researchers can use transposon-based mutagenesis to create random mutations in an organism's genome and then identify the effects of these mutations on gene function.
- ***Complementation assays***: In cases where a gene of interest has been knocked out or mutated, researchers can introduce a functional copy of the gene back into the organism to restore its normal function. This is known as a complementation assay and can help confirm the specific role of the gene.
- ***Gene expression analysis***: Reverse genetics studies often involve analyzing gene expression patterns using techniques like Northern blotting, Western blotting, or quantitative PCR (qPCR). These methods help researchers understand when and where genes are active and how their expression levels change under different conditions.
- ***Phenotypic analysis***: Ultimately, the goal of reverse genetics is to understand how specific genetic changes affect an organism's phenotype. This involves observing and characterizing changes in an organism's appearance, behavior, physiology, or other traits as a result of gene manipulation.

These traditional techniques have paved the way for a deeper understanding of gene function and have been instrumental in fields such as molecular biology, genetics, and developmental biology. However, the disadvantages of traditional reverse genetics techniques are as follows:

- ***Time-consuming***: Traditional techniques, such as homologous recombination for gene knockout or generation of transgenic animals, can be time-consuming. These methods often require several months or even years to complete, making them less suitable for rapid experimentation.
- ***Labor-intensive***: Many traditional methods involve complex laboratory procedures, including extensive breeding of animals, maintaining large colonies, and performing numerous molecular biology experiments. These can be labor-intensive and require specialized skills.
- ***Limited precision***: Traditional techniques may lack precision, leading to off-target effects or incomplete gene knockout/knockdown. This imprecision can make it challenging to attribute observed phenotypic changes specifically to the targeted gene.
- ***Limited scalability***: Traditional methods are often less scalable and less amenable to high-throughput experiments. This limitation can be a significant drawback when studying multiple genes or conducting genome-wide screens.
- ***Low efficiency***: Some traditional methods, such as random transposon mutagenesis, have a low efficiency in generating mutants with the desired phenotype. Researchers may need to screen a large number of individuals to identify the desired mutants.
- ***Species limitations***: Certain traditional techniques are species-specific and may not be easily transferable to other organisms. This can hinder comparative studies and limit the range of organisms in which gene function can be explored.
- ***Difficulty in conditional manipulations***: Temporally and spatially controlled gene manipulations are often challenging with traditional methods. For example, turning a gene on or off at a specific developmental stage or in a specific tissue can be complex.
- ***Risk of genetic compensation***: In gene knockout studies, organisms may compensate for the loss of a specific gene by upregulating other genes or pathways, which can confound the interpretation of phenotypic effects.
- ***Ethical and regulatory concerns***: Traditional methods, especially those involving animals, can raise ethical and regulatory concerns related to animal welfare and genetic modification. These concerns can lead to limitations on the use of these techniques.

- ***Inability to study non-coding elements***: Traditional techniques are often less suitable for studying non-coding regions of the genome, such as regulatory elements and non-coding RNAs, which are now known to play critical roles in gene regulation and function.

Recent Reverse Genetics Techniques

Several recent approaches and technologies have emerged for studying reverse genetics to overcome the limitations of traditional methodology. Here are some notable recent approaches for studying reverse genetics:

- ***Prime editing***: Prime editing is a revolutionary genome editing technique that allows for precise and highly specific changes to the DNA sequence. It offers greater precision than traditional CRISPR/Cas9 by directly rewriting the DNA at the target site. Prime editing has the potential to correct genetic mutations responsible for various diseases.
- ***Base editing***: Base editing is another precise genome editing technique that enables the direct conversion of one DNA base pair into another without causing double-strand breaks in the DNA. It has been used to correct point mutations associated with genetic diseases.
- ***High-throughput functional genomics***: Advances in high-throughput sequencing technologies have facilitated large-scale functional genomics studies. These studies involve the systematic knockout, knockdown, or activation of genes on a genome-wide scale to understand their roles in various biological processes. CRISPR-based screens, such as CRISPR knockout and activation screens, have become powerful tools in this context.
- ***Single-cell genomics***: Single-cell RNA sequencing (scRNA-Seq) has gained prominence for studying gene expression at the individual cell level. It allows researchers to uncover cellular heterogeneity, identify rare cell types, and understand cell lineage and differentiation processes.
- ***Metagenomics and functional metagenomics***: These approaches are used to study the genetic content and functional potential of microbial communities. Metagenomics involves sequencing the collective DNA of an environmental sample, while functional metagenomics explores the activity of genes in those communities.
- ***Synthetic biology and gene circuits***: Synthetic biology techniques enable the design and construction of novel genetic circuits and biological systems. Researchers can engineer cells to perform specific functions, respond to environmental cues, and create custom-designed organisms.

- ***CRISPR/Cas systems beyond Cas9***: While CRISPR/Cas9 was the first widely adopted genome editing tool, newer Cas proteins, such as Cas12 and Cas13, have been developed for different applications. For instance, Cas12 can be used for precise DNA cleavage, while Cas13 targets RNA for editing.
- ***Epigenome editing***: Epigenome editing techniques allow researchers to modify epigenetic marks (e.g., DNA methylation or histone modifications) at specific genomic loci. This can provide insights into the regulation of gene expression and epigenetic contributions to disease.
- ***In vivo gene therapy***: In the context of reverse genetics in humans, in vivo gene therapy approaches have advanced significantly. Techniques like CRISPR-based genome editing can be applied directly to target tissues or cells in living organisms to treat genetic diseases.
- ***Advanced imaging technologies***: High-resolution imaging techniques, such as super-resolution microscopy and live-cell imaging, have improved our ability to visualize cellular processes and dynamic changes in gene expression (Flowchart 2).

Flowchart 2: The stepwise processes involved in both *A*, traditional reverse genetics technique and *B*, recent reverse genetics techniques showcasing the transition from conventional methods to more advanced and data-driven approaches. (CRISPR-Cas9: clustered regularly interspaced short palindromic repeats–associated protein 9; PCR: polymerase chain reaction; RNA-Seq: RNA sequencing)

Conclusion

The field of genetics has evolved from Gregor Mendel's pioneering work to a modern era of profound discoveries and technological advancements. While traditional forward genetics and reverse genetics techniques have been fundamental, recent innovations in genomics, genome editing, and functional genomics have revolutionized our ability to explore the genetic code and gene function with unprecedented precision and efficiency. These contemporary approaches offer powerful tools for unraveling the complexities of genes and their roles in diverse biological processes (Gurumurthy *et al.*, 2016). They have propelled genetics into new frontiers, accelerating progress in medicine, agriculture, and evolutionary biology. As we continue to navigate this genetic landscape, the integration of traditional wisdom with cutting-edge technologies promises to unlock further secrets of life's genetic mysteries.

References

Batel, D.P. "MicroRNAs: Genomics, Biogenesis, Mechanism and Function." Cell 116 (2004): 281-97.

Cheema, J.D., Dicks J. "Computational Approaches and Software Tools for Genetic Linkage Map Estimation in Plants". Briefings in Bioinformatics 10, 6 (2009):595-608.

Drenkard, E. "A Simple Procedure for the Analysis of Single Nucleotide Polymorphisms Facilitates Matp-based Cloning in Arabidopsis." Plant Physiol 124 (2000):1483-92.

Galvão, V.C., Nordström, K.J., Lanz, C., Sulz, P., Mathieu, J., Posé, D., et al. "Synteny-based Mapping-by-Sequencing Enabled by Targeted Enrichment." *The Plant Journal* 71, 3 (2012):517-26.

Gurumurthy, C.B., Grati M.H., Ohtsuka M., Schilit S.L., Quadros R.M., Liu, X.Z. "CRISPR: A Versatile Tool for both Forward and Reverse Genetics Research." *Human Genetics* 135 (2016):971-6.

Jiang, F., Taylor D.W., Thompson A.J., Nagales E., Doudna J.A. "Structures of a CRISPR-Cas9R Loop Complex Primed for DNA Cleavage." *Science* 351, 6275 (2016):867-71.

Kapranov, P., Willingham A.T., Gingeras T.R. "Genome-wide Transcription and the Implications for Genomic Organization." *Nature Reviews Genetics* 8, 6 (2007):413-23.

Lindner, H., Raissig M.T., Sailer C., Shimosato-Asano H., Bruggmann R., Grossniklaus U. "SNP-Ratio Mapping (SRM): Identifying Lethal Alleles and Mutations in Complex Genetic Backgrounds by Next-generation Sequencing." *Genetics* 191, 4 (2012):1381-6.

Mortazavi, A., Williams B.A., McCue K., Schaeffer L., Wold B. "Mapping and Quantifying Mammalian Transcriptomes by RNA-Seq." *Nature Methods* 5 (2008):621-8.

Peters, J.L., Cnudde F., Gerats T. "Forward Genetics and Map-based Cloning Approaches." *Trends in Plant Science* 8, 10 (2003):484-91.

Raingam, M., Bhatt M., Sinha A., Lohani P. 2018. Forward and reverse genetics in plant breeding. *Advanced Molecular Plant Breeding: Meeting the Challenge of Food Security.* Boca Raton, Florida: CRC Press. pp. 49-76.

Schneeberger, K. "Using Next-generation Sequencing to Isolate Mutant Genes from Forward Genetic Screens." *Nature Reviews Genetics* 15, 10 (2014): 662.

12

Pre-breeding: Its Application in Plant Breeding

Raiza Christina G.[1], Sumaiya Sulthana J.[2], Puja Mandal[3] and Sivabharathi R. C.[4]

[1]Ph.D. Scholar, Department of Plant Breeding and Genetics, Agricultural College and Research Institute, TNAU Madurai, Tamil Nadu, India

[2-4]Ph.D. Scholar, Department of Genetics and Plant Breeding, School of Post Graduate Studies, TNAU, Coimbatore, Tamil Nadu, India

Abstract

Pre-breeding, an essential precursor to formal breeding programs, involves harnessing the genetic diversity of wild and underutilized plant species to enhance cultivated crops. Through meticulous selection, hybridization, and backcrossing, pre-breeding introduces valuable traits such as disease resistance, tolerance to environmental stresses, and improved nutritional content into domesticated crops. This process not only broadens the genetic base of cultivated varieties but also equips crops with the resilience needed to address emerging challenges in agriculture, such as changing climates and evolving pests. As a cornerstone of crop improvement, pre-breeding contributes to sustainable agriculture by conserving genetic resources, reducing chemical inputs, and laying the groundwork for the development of novel, adaptable, and productive crop varieties (Cooper et al. 2001).

Keywords: *Pre-breeding, genetic resources, crop improvement*

Introduction

Crop improvement lies at the heart of ensuring food security, agricultural sustainability, and the resilience of our farming systems. As we confront a world characterized by climate variability, emerging pests, and a growing global population, the need for innovative approaches to crop enhancement becomes paramount. This is where the concept of pre-breeding steps in: a fundamental phase that acts as a bridge between the untamed genetic diversity of wild plants and the refined traits of cultivated crops. Pre-breeding represents the exploration, identification, and incorporation of valuable genetic traits from the vast array of plant species that often remain on the fringes of agricultural attention. These "wild relatives" and underutilized varieties possess traits that have evolved over time to withstand diverse environmental stresses, combat diseases, and adapt to changing conditions. Pre-breeding aims to tap into these reservoirs of genetic potential, merging them with the qualities we value in our cultivated crops. At its core, pre-breeding is a scientific journey of discovery and innovation. It involves the strategic selection and hybridization of plant materials, guided by a vision of enhancing crop performance across multiple fronts. The goal is to create crop varieties that are not only productive and high-yielding, but also equipped to thrive in the face of evolving challenges. This involves a delicate dance between tradition and cutting-edge techniques as well as collaboration between breeders, geneticists, and researchers (Qureshi *et al.* 2014).

The process begins with the exploration and collection of genetic material from wild and underutilized sources, encompassing the remarkable diversity that nature offers. This genetic treasure trove holds the potential to bolster our crop's resistance to diseases, improve their water and nutrient efficiency, and enable them to flourish in a range of climates. Through meticulous breeding and selection, pre-breeding introduces these valuable traits into cultivated crops, often through multiple generations of hybridization and careful selection. Pre-breeding is more than just an academic exercise; it is a forward-looking strategy with real-world impact. It equips our crops to adapt the uncertainties of a changing environment, reduces our reliance on chemical interventions, and addresses the nutritional needs of a growing global population. Moreover, pre-breeding aligns with the principles of conservation and sustainable use of genetic resources, ensuring that the genetic diversity of our plant species remains intact for future generations. As we embark on this journey, we uncover the potential to revolutionize the way we approach agriculture. Pre-breeding offers a glimpse into a future where crops are not merely products of cultivation, but co-creations with the natural world. It is a testament to our ability to merge scientific understanding with the wisdom of nature, resulting in crops that are not only resilient but also respect the delicate balance of our ecosystems (Ceccarelli and Grando 2007).

In the chapters that follow, we delve deeper into the strategies, methodologies, and real-world applications of pre-breeding in crop improvement. We explore how this stage of innovation shapes the trajectory of agriculture, from the field to our tables, and how it contributes to a more sustainable, adaptable, and prosperous agricultural landscape.

What is Pre-breeding?

According to the Food and Agriculture Organization of the United Nations (FAO, 1996), pre-breeding is the transfer of beneficial genes from exotic or wild (unadapted sources) types into agronomically appropriate background or breeding material. Pre-breeding is also known as "the art of identifying desired traits and incorporating these into modern breeding materials", according to the Global Crop Diversity Trust. Pre-breeding, to put it simply, is the introduction of genes or gene combinations from sources that have not undergone adaptation into breeding materials, including those that, although adapted, have undergone any type of selection for improvement. Pre-breeding and the results of it must be valuable in order to be incorporated into regular breeding programs. Hallauer and Filho (1981) consider that exotics for pre-breeding purposes encompass any germplasm that does not have immediate value without selection for adaptation to a specific environment. This is despite the fact that there are many notions of exotics. In this respect, races, populations, inbred lines, etc., serve as representations of foreign germplasms. As a result, the products of crosses between adapted and exotic materials—where various amounts of introgression are discovered and assessed,—have been referred to as semi-exotic materials. Pre-breeding aims to increase the genetic diversity of the germplasm, and the enhanced germplasm is easily applicable to routine breeding programs for cultivar improvement.

Germplasm

Germplasm refers to the collection of genetic material, such as seeds, plant tissues, pollen, or other reproductive materials, that can be used to propagate and preserve the genetic diversity of a particular species. It contains the genetic information that determines the traits and characteristics of an organism and plays a crucial role in developing new and improved varieties of crops as well as in preserving the genetic diversity of plant species. Germplasm plays a vital role in improving crop cultivars utilizing the useful traits.

Exotic Lines

Exotic lines do not have an immediate usefulness without selection for adaptation. Generally, the direct use of exotic lines limits the breeding activities in the breeding programs due to high possibility for the introduction of inferior alleles, which negatively affects adaptedness and linkage of undesirable genes with desirable traits (linkage drag).

Gene Pool

In 1971, Harlan and deWet put out the idea of a gene pool. All of an individual's genes and their alleles make up their gene pool. The three groupings of genes make up the gene pool, which was augmented by genetic engineering. These are discussed as follows:

1. ***Primary gene pool*:** This comprises the crop's wild and domesticated relatives. Through standard plant breeding techniques, crossing between group members can result in normal seed set, segregation, and recombination.
2. ***Secondary gene pool*:** Meiotic abnormalities result in hybrids that are partly sterile when the primary gene pool and secondary gene pool are combined.
3. ***Tertiary gene pool*:** Crossing between the gene pools results in infertile offspring. Specialized methods like embryo rescue, tissue culture, chromosomal doubling, and bridging species might get over these crossability obstacles.
4. ***Quaternary gene pool*:** This gene pool is often referred to as the gene ocean. Through the use of recombinant DNA technology, genes may be transmitted not only across species or genera of the same creature but also between distinct organisms.

Why Pre-breeding is Required?

Any crop development program's effectiveness depends on the availability of enough genetic diversity, but this diversity must be in a form that can be used in traditional agriculture. Progress in breeding is hampered by a lack of variation. Food security clearly faces a danger from a limited genetic base. In agroecosystems, genetically homogenous current varieties are replacing genetically varied native cultivars and landraces. Increased genetic susceptibility to illnesses and pests was a result of genetic homogeneity. For greater climatic adaption, look for novel genes and characteristics. Pest and disease populations are changing, encouraging plant breeders to search gene banks for fresh sources of resistance. The most effective method of tying breeding projects and genetic resources together is pre-breeding. Pre-breeding is now a regular, required component of all plant breeding efforts and germplasm diversification initiatives that are planned in advance.

A second crucial goal of genetic modification is to increase yield levels to unprecedented heights. Although this objective is more frequently aspired for than realized, it is true that the lineage of the majority of breakthrough cultivars is very heterogeneous. Examples include semi-dwarf wheat, high-yield dwarf rice, hybrid sorghums, and even maize varieties. In each instance, the creation of the ground-breaking, high-yield cultivars was preceded by considerable pre-breeding. Various types of germplasm were adapted to new genetic environments and geographical locations through pre-breeding. To introduce new, superior features that are not present in the indigenous cultivars, genetic enhancement is utilized. Examples include new percentages of protein in wheat or unexpected starch characteristics in maize. In order to use the genetic diversity, resulting from wild relatives and other unimproved materials in crop improvement programs, pre-breeding is the first step in connecting that variability. The plant breeder and the germplasm curator must work together to comprehend the extent and importance of germplasm collections as well as how new features from these collections might be developed into new varieties. Pre-breeding is chosen based on the anticipated effectiveness, efficiency, and success of eventually transferring the goal characteristics into cultivars for farmers and the source of a desired gene(s). Pre-breeding is utilized, if desirable genes are not present in gene bank accessions that are well suited to the target environment, closely related wild species that can be crossed with crop species, and more distant wild species that are more challenging to cross (Jain and Omprakash 2019).

Objectives of Pre-breeding

The process of pre-breeding often focuses on the following breeding goals:

- Better germplasm and related genetic information that increase the expression and variety of resistance
- Using a larger pool of genetic material to improve yield, pest and disease resistance, and other quality attributes without reducing genetic uniformity in crops.
- Determining desired qualities or genes and then transferring them to an appropriate group of parents to be used in future selection
- Better selection techniques and enhanced parental stocks that are easily used in breeding programs
- Locate potentially valuable genes in a gene bank that is organized and well-documented
- Creating plans that result in the improvement of germplasm to be prepared for use in varietal development

Strategies for Pre-breeding

Pre-breeding strategies involve various methods and techniques to harness the genetic diversity present in wild and underutilized plant populations and transfer valuable traits into cultivated crops. These strategies lay the foundation for successful crop improvement programs. Some common pre-breeding strategies are mentioned as follows (Fig. 1):

- ***Germplasm collection and characterization***: Collect diverse genetic material (germplasm) from wild relatives, landraces, and underutilized varieties. This material is then characterized for various traits of interest, both morphological and genetic, to identify potential sources of valuable traits.
- ***Trait mapping and identification***: Utilize techniques such as molecular markers and genotyping to identify specific genetic regions associated with desirable traits in wild populations. This helps streamline the selection process for these traits during breeding (Tanksley and McCouch 1997).
- ***Introgression and backcrossing***: Transfer target traits from wild or related species into cultivated crops through controlled crossbreeding and backcrossing. This process involves repeatedly crossing the hybrid with the cultivated parent to recover the desired traits while minimizing the genetic contribution from the wild relative.
- ***Pyramiding***: Combine multiple valuable traits into a single plant through repeated crossing and selection. Pyramiding enhances the resilience and adaptability of cultivated crops by accumulating multiple layers of genetic protection against various stresses.
- ***Resynthesis of cultivars***: Reconstruct cultivated varieties by crossing them with their wild ancestors or closely related species. This strategy can reintroduce lost genetic diversity while maintaining the desired agronomic traits.

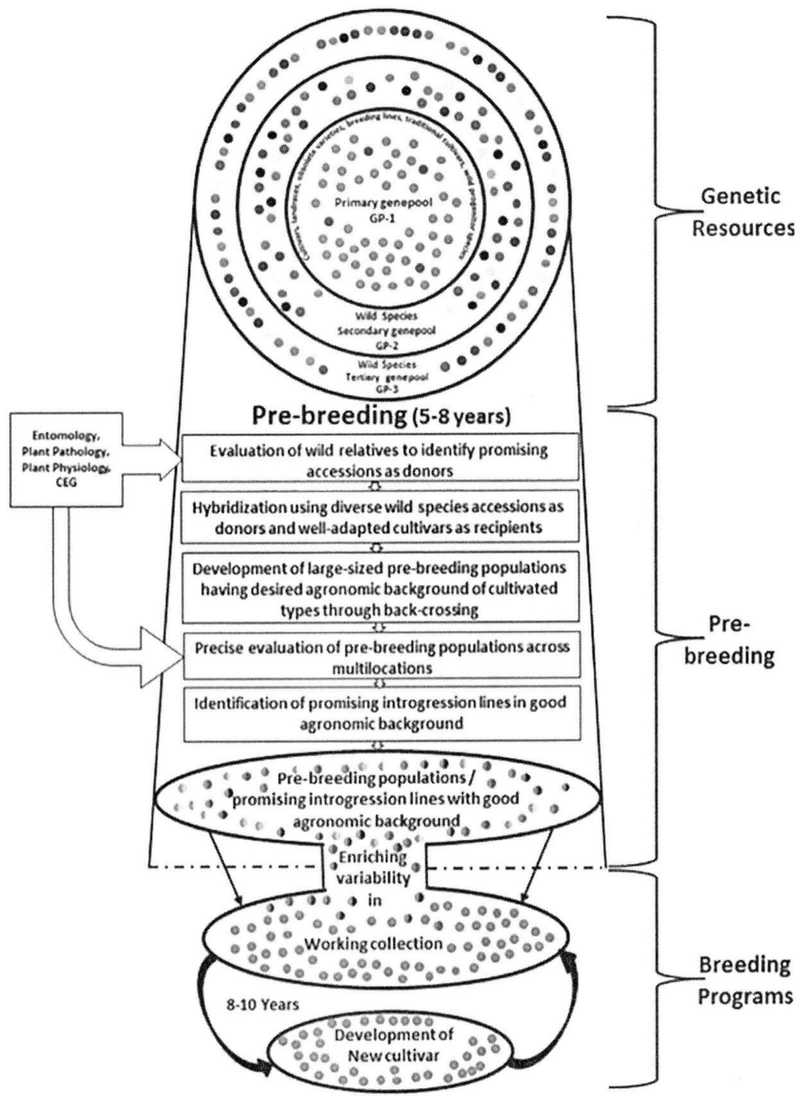

Fig. 1: Pre-breeding as a bridge between genetic resources and crop improvement. (CEG: Center of Excellence in Genomics)

Source: Sharma *et al.* 2013.

- ***Population improvement***: Create populations with diverse genetic backgrounds and traits through controlled crosses. These populations serve as a source of genetic material for subsequent breeding effort.

- ***Hybridization***: Create hybrids between cultivated crops and wild relatives to introduce novel traits. These hybrids can possess a mix of attributes from both parents and may exhibit improved performance under specific conditions.
- ***Mutation breeding***: Induce genetic mutations in wild or underutilized plants to generate variation and select for desired traits. This approach can accelerate the discovery of useful genetic variants.
- ***In vitro culture and micropropagation***: Use tissue culture techniques to rapidly propagate and multiply plants with desired traits, especially those that are difficult to crossbreed or hybridize.
- ***Genome editing***: Employ advanced techniques like CRISPR-Cas9 (clustered regularly interspaced short palindromic repeats–associated protein 9) to precisely modify the genes of cultivated crops. This can result in the introduction of targeted traits without introducing additional genetic material.
- ***Cisgenesis and transgenesis***: Cisgenesis involves transferring genes from a related species that could also sexually cross with the cultivated crop. Transgenesis involves introducing genes from unrelated species. These approaches can provide access to traits not present in the crop's gene pool.
- ***Genome-wide association studies (GWAS)***: Identify associations between genetic markers and specific traits across diverse germplasm. This allows for the identification of candidate genes responsible for desired traits.
- ***Phenotypic selection***: Evaluate and select individuals based on their physical traits and performance under specific conditions. This strategy is particularly important when genetic markers for desired traits are not readily available.
- ***Participatory breeding***: Involve local farmers, communities, and stakeholders in the pre-breeding process to ensure that the developed varieties meet their specific needs and preferences.
- ***High-throughput screening***: Use automated technologies to rapidly assess a large number of individuals for specific traits. This accelerates the selection process and allows breeders to work with larger populations.
- ***Conservation and sustainable use***: Prioritize the conservation of wild and underutilized crop relatives to ensure the continuous availability of genetic resources for future pre-breeding and breeding efforts (Seetharam 2007).

- *Cross-species hybridization*: Explore the potential for hybridization between different but related species to introduce novel traits that might not exist within the immediate gene pool of the cultivated crop.

Each pre-breeding strategy has its own advantages and challenges, and the choice of strategy depends on factors such as the target crop, the specific traits of interest, available resources, and the timeline for developing new varieties. Successful pre-breeding lays the groundwork for subsequent breeding programs that focus on refining and stabilizing the introduced traits in order to create improved, resilient, and productive crop varieties.

Applications of Pre-breeding

Pre-breeding has four primary applications:

1. Increasing the genetic diversity of an organism to lessen its vulnerability
2. Locating traits in exotic materials and transferring those genes into material that is easier for breeders to access
3. Introducing genes from wild species into breeding populations when doing so appears to be the most efficient tactic
4. Locating novel genes from unrelated species using genetic transformation techniques

Pre-breeding is used to boost access to and use of genetic differences preserved in gene banks, which helps crop development initiatives be more successful and efficient (Kumar and Shukla 2014).

Challenges in Pre-breeding

The challenges of pre-breeding include:

- Scarcity of donors for certain qualities, such as resistance
- Lack of assessment and characterization data
- Awareness of genetic variability
- Effective breeding program and financial sources
- Interspecific cross compatibility issues
- Chromosome pairing and stability barriers in hybrids, which limit the transfer of genes from wild species into domesticated ones
- Linkage drags
- Hybrid sterility and in viability.
- Interspecific hybrid population's small sample size
- Limited genetic recombination in hybrid population

- Challenges in exchanging and gaining access to farmed germplasm owing to legal constraints like intellectual property rights (IPR)

Conclusion

Pre-breeding emerges as a dynamic and transformative strategy within the realm of crop improvement. By tapping into the rich genetic diversity present in wild and underutilized plant populations, pre-breeding infuses cultivated crops with resilience, adaptability, and enhanced traits. As the world faces escalating challenges posed by climate change, disease outbreaks, and resource limitations, pre-breeding stands as a potent tool to bolster global food security. Through meticulous crossbreeding, hybridization, and selection, pre-breeding not only diversifies and strengthens crop gene pools but also aligns with sustainable agricultural practices by reducing reliance on chemicals and conserving vital genetic resources. As the bridge between untamed genetic potential and the exigencies of modern agriculture, pre-breeding showcases the fusion of nature's ingenuity and human innovation, offering a pathway toward a more resilient, productive, and sustainable agricultural future.

References

Ceccarelli, S., Grando S. "Decentralized-Participatory Plant Breeding: An Example of Demand Driven Research." Euphytica 155 (2007):349-60.

Cooper H.D., Spillane C., Hodgkin T. 2001. "Broadening the genetic base of crops: an overview" In Broadening the Genetic Base of Crop Production Wallingford: CABI Publishing in cooperation with FAO and IPGRI, CAB International. pp. 1-23.

Dwivedi, S. L., Upadhyaya, H. D., Stalker, H. T., Blair, M. W., Bertioli, D. J., Nielen, S., & Ortiz, R. (2008). Enhancing crop gene pools with beneficial traits using wild relatives. Plant Breeding Reviews, 30, 179-230.

FAO. 1996. Global Action Plan for the Conservation and Sustainable Utilization of Plant Genetic Resources for Food and Agriculture and the Leipzig Declaration adopted by the International Technical Conference on Plant Genetic Resources Jure Leipzig, Germany. Rome: Food and Agriculture Organization of the United Nations.

Hallauer, A.R., Filho J.B.M. 1981. Quantitative Genetics in Maize Breeding United States: Iowa State University Press.

Haussmann, B.I.G., Parzies, H. K., Presterl, T., Susic, Z., & Miedaner, T. (2004). Plant genetic resources in crop improvement. Plant Genetic Resources, 2(1), 3-21.

Jain S.K., Omprakash. "Pre-breeding: A Bridge between Genetic Resources and Crop Improvement." International Journal of Current Microbiology and Applied Sciences 8, 02 (2019):1998-2007.

Kumar, V., Shukla Y.M. "Pre-breeding: Its Applications in Crop Improvement." Research News for U (RNFU) 16 (2014):2250-3668.

Meena, A.K., Gurjar, D., & Kumhar, B.L. (2017). Pre-breeding is a bridge between wild species and improved genotypes—A review. Chem Sci Rev Lett, 6(22), 1141-1151.

Qureshi, A.M.I., Lone A., Wani A., Wani S.H., Nehvi F. "Pre-breeding and Population Improvement." LS International Journal of Life Sciences 2, 3 (January 2014):188-97.

Seetharam, A. "Pre-breeding: An Important Step in the Effective Utilization of Conserved Germplasm." National Workshop on Utilization of Wild Mulberry Genetic Resources (November 2-3, 2007):9-16.

Shankar, M., Francki, M., & Loughman, R. (2012). Pre-breeding for disease resistance in wheat–The stagonospora nodorum blotch example. Microbiology Australia, 33(1), 6-8.

Sharma S., Upadhyaya H.D., Varshney R.K., Gowda C.L.L. "Pre-breeding for Diversification of Primary Gene Pool and Genetic Enhancement of Grain Legumes" Frontiers in Plant Science 4 (2013): 309.

Tanksley, S.D., McCouch S.R. "Seed Banks and Molecular Maps: Unlocking Genetic Potential from the Wild." Science 277 (1997):1063-6.

13

Role of Wide (Distant) Hybridization in Crop Improvement: Applications and Limitations

Poonam Sharma[1], Uma Bharti[2], Neha Jha[3] and Dharm Veer Singh[4]

[1,2]Ph.D. Scholar, Department of Genetics and Plant Breeding, Chaudhary Sarwan Kumar Himachal Pradesh Krishi Vishwavidyalaya, Palampur, Himachal Pradesh India

[3]Ph.D. Scholar, Department of Genetics and Plant Breeding, Sardar Vallabhbhai Patel University of Agriculture and Technology, Meerut, Uttar Pradesh, India

[4]Dpeartment of Genetics and Plant Breeding, Banda University of Agriculture and Technology, Banda, Uttar Pradesh, India

---

Abstract

Wide hybridization or distant hybridization is crossing between two different species or genera, and has been used successfully to transfer genes and to create new crop species. Wild plant species related to crop plants form an important source of useful traits for quality improvement and biotic and abiotic stress tolerance. Cultivated species has lost many beneficial traits for stress tolerance in the process of domestication and selection, which resulted in uniformity in many agronomic traits. Although there have been multiple cases of valuable tolerant genes being transferred from wild rice to cultivated rice, it is now necessary to look at new breeding and selection strategies that may be applied to wide hybridization. It breaks the species barrier for gene transfer and makes it possible to transfer the genome of one species to another, which results in changes in genotype and phenotype of the progenies. The barriers in wide hybridization occurs as prezygotic and postzygotic barriers. Techniques of chromosome doubling, bridging species, protoplast fusion, and embryo rescue are highly beneficial in recovering fertile progenies by overcoming various barriers in wide crossings. Broadening the gene pool of a crop is an important plant breeding method as it can enhance tolerance of major biotic and abiotic stresses and improve the quality characteristics of the plant. In this chapter, we discuss about barriers related to hybridization between different species, advantages and limitations encountered, and the future prospectives for a successful breeding program.

Keywords: *Abiotic stress, crop improvement, genetic diversity, green revolution, offspring*

Introduction

India went from having a food grain deficit to having a surplus because of the green revolution. The green revolution has had the biggest impact on agricultural development than any other endeavor. Additionally, it has shown how it has affected various crop breeding and their production. Reduced varietal diversity in the most important crop species was one of the side consequences of the green revolution, as was an increase in uniformity in appearance and harvestable goods. This made enhanced agriculture more vulnerable to natural disasters. Disease outbreaks were driven by the emergence of new pathogen races, and pest attacks reduced crop yields by up to 50%. Changing climatic conditions have caused abiotic stresses like drought, flood, salinity, and high temperature, which in turn results in the reduction of yield and quality.

Wide hybridization has been promoted as a powerful tool in the hands of plant breeders in order to restore the characteristics of ecological sustainability and to battle the biotic and abiotic stress in cultivated crops, as wild species are the rich source of noble characters, have better quality and processing traits as well as impart resistance against biotic and abiotic stress. Crop wild relatives have been utilized for breeding for decades, particularly to transfer genes that make cultivated species more resistant or tolerant to pests, diseases, or abiotic stress. Wide hybridization is an excellent approach for crop improvement that combines modern molecular techniques with effective conventional breeding (Gowda *et al.* 2020).

Theory

Wide hybridization as a norm is an attempt to intermate two species of a genus or two genera of a taxon with the intention of introgression of genes of economic value into the cultivated species. Wide hybridization invariably comprises crosses between wild, primitively cultivated species and genera.

Interspecific hybridization: It means hybridization between individuals from different species belonging to the same genus (Fig. 1).

Examples include:

Fig. 1: Interspecific hybridization.

- *Nerica*: An upland rice for Africa
- *Oryza sativa (Asian upland rice)*: Non-shattering, resistant to lodging, high yield potential
- *Oryza glaberrima* (African rice): Drought-tolerant, disease-resistant, weed-suppressing
- Nerica rice combines the best of both species.
- *Brassica* species within the triangle of U; three diploid species (*Brassica rapa*, *Brassica nigra*, and *Brassica oleracea*) which represent the AA, BB, and CC genomes are shown. Also shown are three allotetraploid species (*Brassica juncea*, *Brassica napus*, and *Brassica carinata*) which are hybrid combinations of the basic genomes. Diploid chromosome number (2n) is shown in Figure 2 (Katche *et al.* 2019; Koh *et al.* 2017).

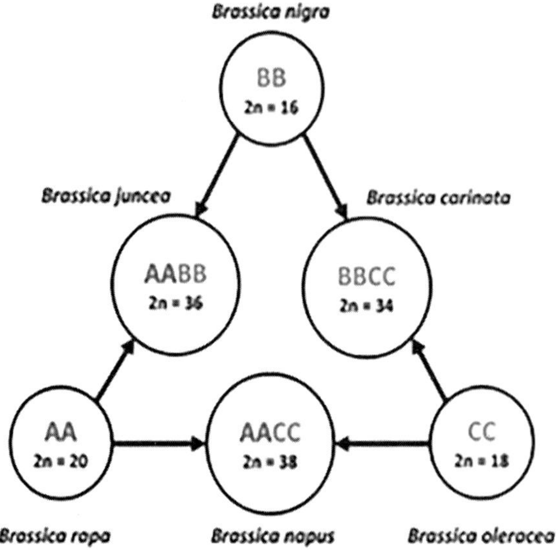

Fig. 2: *Brassica* species within the triangle of U.

- ***Intergeneric hybridization***: It means hybridization between individuals from different genus belonging to same family (Liu *et al.* 2014) (Fig. 3).

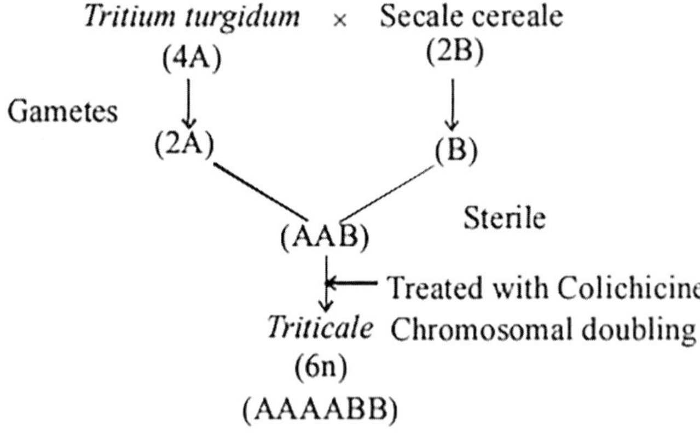

Fig. 3: Intergeneric hybridization.

- *Triticale*: Cereal created in the laboratory
- Triticale, a cross (intergeneric cross) between *wheat and rye*, was produced by embryo rescue of the product of fertilization and a chemically induced doubling of the chromosomes.

Role of Wide Hybridization

The role of wide hybridization covers disease resistance against various pathogens, insect resistance, abiotic stress resistance, quality improvement, yield enhancement, development of new crop species, transfer of sterile cytoplasm for hybrids production, and rootstock breeding.

Barriers of Wide Hybridization

- ***Failure of zygote formation/cross incompatibility***: Inability of the functional pollens of one species or genera to effect fertilization of the female gametes of another species or genera is referred to as cross incompatibility.

  It may be due to:

  - Failure of fertilization because the pollen may not germinate.
  - Pollen tube is unable to reach to embryo sac and hence sperms are not available for fertilization.
  - Pollen tube may burst in the style of another species, e.g., *datura.*
  - The style of the female parent may be longer than the usual length of the pollen tube growth, therefore the pollen does not reach the embryo sac, e.g., *Zea mays* and *Tripsacum spp.*

- ***Failure of zygote development/hybrid inviability***: The inability of a hybrid zygote to grow into a normal embryo under the usual conditions of development is referred to as hybrid inviability.

  This may be due to:

  - Lethal genes
  - Genetic disharmony between the two parental genomes
  - Chromosome elimination
  - Incompatible cytoplasm
  - Endosperm abortion

- ***Failure of hybrid seedling development/hybrid sterility***
- ***Hybrid sterility***: Hybrid sterility refers to the inability of a hybrid to produce viable offspring. The main cause of hybrid sterility is lack of structural homology between the chromosomes of two species.
- Some distant hybrids die during seedling development or even after initiation of flowering. The mechanisms involved in the failure of seedling development most likely involve complementary lethal genes.

***Examples include*:**

- In cotton, certain interspecific hybrids appear normal, but die in various stages of seedling growth; some plants die at flowering.
- Interspecific and intergeneric F_1 hybrids of wheat show both chlorosis and necrosis.

Techniques to Overcome Pre-fertilization Barriers

***Taxonomic position of parental species*:** An ease in hybridization is expected when the species more resemble phenotypically. Taxonomic classification is based on morphological features but by and large is the outcome of genetic factors in association with environment.

***Doubling of chromosome number*:** When the failure of hybridization is due to different ploidy level then this technique is followed. In many polyploid cultivated species, their wild progenitors are diploid and crossing attempts are difficult. One may increase ploidy level of wild type by colchicine treatment. This enhances the success rate in crops like potato, *Cucumis*, and *Brassica* spp.

***Bridging species technique*:** In many crops, two species which are otherwise incompatible, may be hybridized with the help of third species. The third species acts as bridge in recombining the two incompatible species so known as bridging species. This technique has been used in making wide crosses in potato, lettuce, and sugar beet. Hayes *et al.* (2005) were successful in using *Solanum verrucosum* as a female bridging parent to access 2× 1([EBN] endosperm balance number) *Solanum pinnatisectum*. *Solanum simplicifolium* was used as bridging species in crossing *Solanum acaule* and *Solanum tuberosum*.

***Shortening of the style*:** In some species, the incompatible reaction can be overcome by reducing the length of style. Incompatible reactions in radish can also be overcome by the reduction of stylar tissues.

***Mentor pollination*:** Pollen grains of distant species do not germinate on the stigma of cultivated species. However, when these pollen grains are mixed with killed maternal pollen grains, germination of the incompatible pollen grains takes place as in the case of *Cucumis* (Beharav and Cohen 1994). It happens because cell wall proteins of pollen play a pivotal role in pollen–pistil interaction. Pollen killed in ethanol and mixed with incompatible pollens, releases the proteinaceous recognition factors, thereby masking rejection reaction of the recipient stigma. The killed maternal pollen is known as mentor pollen.

***Use of growth regulators (GRs) and immune suppressants*:** GRs enhance the zygotic formation in distant hybrids of many cucurbits, okra, and tomato. Commonly used GRs are IAA, IBA, and GA3. GA3 (75 ppm) application to maternal plant 1 or 2 days before and after pollination led to improved zygote formation, faster pollen tube growth, more embryo survival, and more seed set-in wide crosses. Application of auxin in *Solanum* prevented flower abscission and enhanced fruit set.

***In vitro* fertilization:** It is effective when stigma and style inhibit pollen tube growth and embryo abortion occurs at early stages of development. The whole gynoecium is excised and placed on Murashige and Skoog (MS) medium followed by dusting of pollens on the stigma and fertilized ovule is reared to maturity (Ondrej *et al.* 2002).

***Protoplast fusion*:** It involves the fusion of protoplast of two incompatible species or genera and the fused heterokaryon is placed on artificial nutrient medium and regenerated into hybrid plant. This approach has particularly more potential to generate new genotypes in vegetatively propagating species. Some examples are tomato + potato and *Solanum nigrum* + *S. tuberosum*. Chandel *et al.* (2015) studied interspecific potato somatic hybrids between cultivated *S. tuberosum* dihaploid C-13 and wild species *Solanum cardiophyllum* via protoplast fusion. The use of somatic fusion for improving valuable agronomic traits in cultivated potatoes is done for traits like atrazine resistance, frost tolerance, quality improvement, and pest and diseases resistance.

Techniques to Overcome Post-fertilization Barriers

***Embryo rescue*:** Embryo abortion is the major barrier in distant hybridization. It can occur at any stage of development depending on the genomic relationship of two parental species. *Solanum sitiens* is a rare endemic plant of the Atacama Desert of Chile having tolerance to drought, salinity, and low temperatures, resistances to certain pathogens, and modified fruit ripening. To overcome sterility and unilateral incompatibility of *Solanum lycopersicum* × *S. sitiens* hybrids, embryo culture was used (Chetelat 2016). *Capsicum chinense*, *Capsicum annuum* and *Capsicum frutescens* were crossed with each other and embryo rescue was done between 27 and 33 days after pollination by Debbarama *et al.* (2013). Hybrid plants were obtained and their hybridity was confirmed using both morphological and RAPD (randomly amplified polymorphic DNA) markers.

***Ovary culture*:** In some wide crosses, embryo abortion occurs at early stages of development when it is difficult to excise and culture embryos. To overcome this problem, ovaries are cultured. Depending on cross combination, ovaries

2–15 days after pollination are excised and cultured on nutrient medium. Flower is pruned by removing calyx, corolla, and stamen. Distal part of the pedicel is cut and ovary planted on simple semi-solid MS nutrient medium. When the embryos become visible, they are aseptically taken out and cultured in a manner of embryo rescue technique. Ovule culture is an elegant experimental system by which ovules are aseptically isolated from the ovary and are grown aseptically on chemically defined nutrient medium under controlled conditions. It is used when barrier impedes growth of zygote at earlier stages of development.

***Backcrossing*:** Wide crosses showing poor fertility can be backcrossed to cultivated species to improve the fertility of hybrids. Backcrossing can be applied to balance cytoplasmic interaction by producing cybrids or alloplasmic lines. Chamola *et al.* (2015) used cytoplasmic male sterile (CMS) lines of *B. juncea* and *B. napus* with the mitochondrial genome of *Moricandia arvensis* and *Erucastrum canariense*, respectively, which were used to transfer CMS to cauliflower (*B. oleracea*). Embryo rescue was also done in BC1 and BC2 to obtain progenies. Recovery of the recurrent parent phenotype was faster in *B. napus* × *B. oleracea* than *B. juncea* × *B. oleracea*. BC3 generation plants of *B. napus* × *B. oleracea* showed good curd formation and complete male sterility and nine bivalents at meiosis whereas those of *B. juncea* × *B. oleracea* were male sterile but still had genetic elements of *B. juncea*.

Table 1: Interspecific versus intergeneric hybridization Singh (2015).

Particulars	**Interspecific hybridization**	**Intergeneric hybridization**
Parents involved	Involve two different species of the same genus	Involve two different genera of the same family
Fertility	Such hybrids vary from completely fertile to completely sterile	Such hybrids always sterile
Use in crop improvement	Frequently used	Less than interspecific crosses
Release of hybrid varieties	Possible in some crops	Not possible
Evolution of new crops	Not possible, but evolution of new species is sometimes possible	Sometimes possible

Applications in Crop Improvement

- ***Alien addition lines*:** Carries one chromosome pair from a different species in addition to somatic chromosome complement. The purpose of the alien additions generally is the transfer of disease resistance from related species.

Example: Transfer of disease resistance from *Nicotiana glutinosa* to *Nicotiana tabacum*

- ***Alien substitution lines***: It has one chromosome pair from different species in place of the chromosome pair of the recipient parent.
- ***Introgression of genes***: Transfer of small chromosome segments with desirable genes.
- Disease and insect resistance (Table 2) Singh (2015).

Table 2: Disease and insect resistance.

Crop	Character transferred	Crop species	Species
Okra	Resistance to YVMV (yellow vein mosaic virus)	*Abelmoschus caillei*	*Abelmoschus esculenta*
Brinjal	Bacterial wilt	*Solanum stenotomum*	*Solanum melongena*
Tomato	*Fusarium* wilt	*Solanum hirsutum*	*Solanum lycopersicum*
Chili	Fruit rot	*Capsicum chinense*	*Capsicum annuum*
Onion	Purple blotch	*Allium fistulosum*	*Allium cepa*
Potato	Late blight, leaf roll, virus X	*Solanum demissum*	*Solanum tuberosum*
Tomato	White fly Root-knot nematode	*S. hirsutum* *Solanum peruvianum*	*Solanum esculentum* *S. esculentum*
Potato	Nematode	*Solanum vernei*	*S. tuberosum*
Okra	Fruit and shoot borer	*Abelmoschus manihot*	*Abelmoschus esculentus*

- Quality improvement in different crops (Table 3) Singh (2015).

Table 3: Quality improvement in different crops.

Crop species	Character transferred	Transferred from species	Transferred to species
Tomato	Carotenoid content	*Solanum hirsutum*	*Solanum esculentum*
	Soluble solid	*Solanum chmielewskii*	*S. esculentum*
Melon	Thick rind and good keeping quality	*Cucumis melo var. cantaloupensis*	*C. melo*
Potato	Starch content	*Solanum acaule*	*Solanum tuberosum*

Example: Triticale

- **Utilization as new hybrid varieties**

Example: F_1 hybrids in cotton (*Gossypium hirsutum* × *Gossypium barbadense*)

Sugarcane: All the present-day commercial varieties are complex interspecific hybrids involving *Saccharum officinarum* and *Saccharum spontaneum*

Limitations of Distant Hybridization

- Incompatible crosses
- F_1 sterility
- Problems in creating new species
- Lack of homoeology between chromosome of the parental species
- Undesirable linkages
- Problems in the transfer of recessive oligo genes and quantitative traits
- Lack of flowering in F_1
- Problems in using improved varieties in distant hybridization
- Dormancy

Achievements

Hybrid Varieties

- ***Upland cotton***: MCU-2, MCU-5, Khandwa 1, Khandwa 2 etc., are derivatives of interspecific hybridization.
- Hybrid between Pearl millet × Napier grass—hybrid Napier is very popular for its high fodder yield and fodder quality, e.g., Jaywant and Yashwant
- ***Interspecific hybrids in cotton***: Varalaxmi, Savitri, DCH-32, NHB-12, DH-7, DH-9, etc.
- Parbhani Kranti variety of okra

Future Prospects

The ultimate goal of wide hybridization programs is to discover beneficial genes and traits hidden in the plant genome and use these discoveries to crop breeding. Crosses between *O. sativa* and wild relatives may result in the finding of beneficial quantitative trait loci (QTLs) from a considerably wider range of allelic variants than that found in cultivated lines (Ashihikari and Sakamoto 2008). With the developing infrastructure of plant genomics, genes coding for key agronomic features can be identified. These scientific breakthroughs and instruments will result in more efficient and practical breeding programs. New technologies that aid in boosting and directing the natural process of meiotic recombination will also facilitate the introgression of beneficial alleles (McCouch *et al.* 2007) and reduce the existing need for massive population screening. Future strategies that combine the use of new tools and procedures with traditional and efficient plant breeding methods, as well as a broad understanding of and access to modern technologies, may allow populations to restore lost characteristics. Also, through wide hybridization, it is possible to replace damaged alleles with functioning copies from allied species.

Conclusion

Finally, wild species provide a storehouse of novel genes that can be used for wide hybridization. As a result, wild species are at the core of a wide hybridization initiative. It has been used effectively in row crops and vegetables. In various agriculture crops, it has been used due to its ability to create novelty. Hybridization can also play a more diverse role in promoting speciation. It may provide the raw material for adaptive divergence or initiate new hybrid populations, potentially leading to speciation. The barrier to wide hybridization includes external (e.g., spatial isolation, ecological isolation, mechanical isolation, and cross incompatibility) and internal (hybrid inviability, hybrid sterility, and hybrid breakdown). Wide hybridization is a powerful strategy for genetic improvement in agricultural crops. It is an extremely technological and knowledge-intensive procedure. Despite the fact that there are numerous challenges to extensive hybridization, they can be overcome using both conventional and biotechnology approaches. Wide hybridization using in vitro and biotechnological approaches is being employed to broaden the genetic base. It can be utilized for disease and pest resistance breeding as well as cultivar quality improvement, which is an essential necessity. Through the display of hybrid vigor, hybridization can influence phenotypic consequences. On longer evolutionary time scales, hybridization can result in local adaptation via the introgression of novel alleles and transgressive segregation as well as the development of new hybrid species in some situations. To feed the ever-increasing population and to fight with malnutrition, wild species offers the scope for quality improvement in various cultivars.

References

Ashihikari, M., Sakamoto T. 2008. "Rice Yielding and Plant Hormones" Biotechnology in Agriculture and Forestry: Rice Biology in the Genomics Era. Edited by Hirano H.Y., Sano Y., Hirai A., Sasaki T. Germany: Springer Berlin Heidelberg. pp. 309-19.

Beharav, A., Cohen, Y. "The Crossability of Cucumis melo and C. metuliferus, an Investigation of In Vivo Pollen Tube Growth." Cucurbit Genetics Cooperative Report 17 (1994): 97-100.

Chamola, R., Balyan H. S., Bhat S. "Transfer of cytoplasmic male sterility from alloplasmic Brassica juncea and B. napus to cauliflower (B. oleracea var. botrytis) through interspecific hybridization and embryo culture." Indian Journal of Genetics and Plant Breeding 73 (2015):203.

Chandel, P., Tiwari J.K., Ali N., Devi S., Sharma S., Sharma S., *et al.* "Interspecific Potato Somatic Hybrids between *Solanum tuberosum* and *S. cardiophyllum*, Potential Source of Late Blight Resistance Breeding." *Plant Cell, Tissue and Organ Culture* 123 (2015):579-89.

Chetelat, R.T. "Overcoming Sterility and Unilateral Incompatibility of *Solanum lycopersicum* × *S. sitiens* Hybrids." *Euphytica* 207 (2016):1-4.

Gowda, P.P., Rafeekher M., Nithinkumar K.R. "Wide Hybridization in Vegetable Crops." *International Journal of Current Microbiology and Applied Sciences* 9 (2020):1025-34.

Hayes, R.J., Dinu I.I., Thill C.A. "Unilateral and Bilateral Hybridization Barriers in Inter Series Crosses of 4x 2EBN *Solanum stoloniferum*, *S. pinnatisectum*, *S. cardiophyllum*, and 2x 2EBN *S. tuberosum* Haploids and Haploid-species Hybrids." *Sexual Plant Reproduction* 17 (2005):303.

Katche, E., Quezada-Martinez D., Katche E.I., Vasquez-Teuber P., Mason A.S. "Interspecific Hybridization for *Brassica* Crop Improvement." *Crop Breeding, Genetics and Genomics* 1 (2019):e190007.

Koh J.C.O., Barbulescu D.M., Norton S., Redden B., Salisbury P.A., Kaur S., *et al.* "A Multiplex PCR for Rapid Identification of Brassica Species in the Triangle of U." *Plant Methods* 13 (2017):49.

Liu, D., Zhang H., Zhang L., Yuan Z., Hao M., Youliang Y. 2014. "Distant Hybridization: A Tool for Interspecific Manipulation of Chromosomes" *Alien Gene Transfer in Crop Plants,* volume 1. Edited by Pratap A., Kumar, J. New York: Springer-Verlag. pp. 25-42.

McCouch, S.R., Sweeney M., Li J., Jiang H., Thomson M., Septiningsih E., *et al.* "Through the Genetic Bottleneck: *O. rufipogon* as a Source of Trait-enhancing Alleles for *O. sativa*." *Euphytica* 154 (2017):317-39.

Ondrej, V., Navrátilová B., Lebeda A. "In Vitro Cultivation of *Cucumis sativus* Ovules after Fertilization." *Acta Horticulturae* 588 (2002):339-43.

Singh, P. (2015). Essentials of Plant Breeding. Sixth edition. Kalyani Publishers.

14

Speed Breeding: Importance and Future Prospects in Agriculture

***Prakash*[1], *Gandikota Brahmani*[2], *Sandhya Lamichaney*[3], *Raman Choudhary*[4] *and Dharm Veer Singh*[5]**

[1]*Ph.D. Research Scholar, Division of Vegetable Science, ICAR-IARI Outreach Campus Indian Institute of Horticultural Research, Bengaluru, Karnataka, India*

[2]*M.Sc. Research Scholar, Department of Vegetable Science, Punjab Agricultural university, Ludhiana, Punjab, India*

[3]*Ph.D. Research Scholar, School of Life Sciences, Department of Horticulture, Sikkim University, Sikkim, India*

[4]*Ph.D. Research Scholar, Department of Silviculture and Agroforestry, Rani Lakshmibai Central Agricultural University, Jhansi, Uttar Pradesh, India*

[5]*Department of Genetics and Plant Breeding, Banda University of Agriculture and Technology, Banda, Uttar Pradesh, India*

Abstract

Conventional crop breeding requires a substantial investment of time, space, and resources for the selection process and the subsequent crossing of desired plants. The time taken for the complete seed-to-seed cycle is a critical constraint in the advancement of plant research and breeding. The concept of speed breeding holds the promise of diminishing the duration necessary for the entire process of cultivar development, from creation to introduction and eventual commercial availability, thereby enhancing the improvement of both food and industrial crops. It revolves around optimizing conditions such as light intensity, temperature, and daylight duration (22-hour light, 22°C day/17°C night, and high light intensity). This approach enhances photosynthesis, triggering early flowering, and, in combination with yearly seed collection. The adoption of speed breeding shows promises in realizing nutritional security and promoting sustainable agriculture for the future rising population. The technique's integration with advanced genomics and gene editing opens doors to precise trait incorporation. Moreover, speed breeding reduces resource consumption, aligning with sustainability goals. As local adaptations become feasible, it empowers diverse regions.

Collaboration in refining these techniques will foster global agricultural innovation, particularly benefiting developing nations. By preserving genetic diversity and swiftly responding to emerging challenges, speed breeding emerges as a transformative force in shaping resilient, high-yielding, and climate-resilient crops. Its potential to revolutionize crop improvement and sustainable agriculture makes it a key player in shaping a more resilient and productive agricultural future.

Keywords: *Speed breeding, crop improvement, genetic diversity, food security, sustainability*

Introduction

As per the United Nations, the global population is projected to reach around 9.9 billion by 2050 (Anonymous 2022). Meeting the needs of this growing population requires the development of innovative technologies to expedite crop production cycles and create high-yielding varieties. However, traditional methods for enhancing crops are time-consuming, necessitating a faster approach to crop breeding to address these challenges. To address this, molecular markers were introduced in the 1990s, allowing for the targeted selection of desirable traits within a single growing season. This significantly reduces the time needed to evaluate genotypes for specific traits. Nonetheless, utilizing molecular markers' demands specialized expertise in marker design and handling, and they also come with a high cost. Alternative techniques include the utilization of winter nurseries, doubled haploids, genetic engineering, and genome editing to shorten the crop breeding cycle. However, these methods have their own drawbacks. Winter nurseries can be expensive and challenging to manage, lacking guarantees of quality seed production. Doubled haploids are unavailable for numerous crops, often requiring skilled personnel and substantial financial resources. Additionally, transgenic or genome-edited crops may not always be viable due to political regulations or societal reservations (Hickey *et al.* 2017, Ghosh *et al.* 2018).

To address these limitations, Lee Hickey and his colleagues introduced an innovative technique called "speed breeding". This technology aims to reduce the breeding cycle duration and hasten crop enhancement through expedited generation cycling. Speed breeding represents a pioneering and advanced agricultural method wherein optimal plant growth conditions are fine-tuned to greatly accelerate the process of breeding and selecting crops. This methodology involves the utilization of controlled environments, including growth chambers or greenhouses, in conjunction with specialized lighting and other parameters. These factors are carefully managed to establish ideal circumstances for rapid plant growth and progression. The central objective of speed breeding revolves around minimizing the time needed to cultivate multiple plant generations within a single year. This ultimately facilitates the swift creation of novel crop varieties. In contrast to traditional breeding timelines, speed breeding achieves this by crafting an environment that spurs plants to grow at an accelerated rate and achieve flowering more promptly. Though manipulating elements such as light intensity, photoperiod variations, temperature levels and nutrient availability, scientists can stimulate plants to attain maturity and generate seeds within notably abbreviated timelines (Ahmar *et al.* 2020; Wanga *et al.* 2021; Samantara *et al.* 2022).

Strategies of Speed Breeding

Manipulation of Photoperiod (Light Duration and Intensity)

Photoperiod pertains to the duration of daily exposure that plants receive under designated light and dark schedules, designed to expedite rapid growth, development, flowering, and seed production (Vince-Prue 1994). Various crop species and different genotypes within those species exhibit distinct photoperiodic requirements for inducing flowering and promoting seed set (Kouressy *et al.* 2008; Saito *et al.* 2009). As a result, it becomes crucial to ascertain the optimal combinations of light quality, intensity, and photoperiod that effectively initiate flowering in diverse crops and genotypes. Modifying the photoperiod can be achieved in a cost-effective manner by utilizing low-energy light-emitting diodes (LEDs), which can be powered by battery-based inverters connected to solar panels. Implementing solar power systems offers a practical and sustainable solution for indoor speed breeding, particularly in regions with an unstable or unreliable electricity supply (Wanga *et al.* 2021).

Temperature Regulation

Modifications in both soil and air temperatures exert a direct influence on germination and growth behaviors, contributing to accelerated progress in growth, flowering, seed production, and maturation. Extreme temperature conditions, whether excessively low or high, trigger a diverse array of effects on the pace of plant development, including a shift from the vegetative to the reproductive stages (Hatfield and Prueger, 2015; McClung *et al.* 2016). The temperature range conducive to germination in most crops typically falls within 12°–30°, whereas the optimal temperature for growth, flowering, and seed production typically ranges from 25° to 30° for the majority of crops. Photoperiod-sensitive genotypes of different crops exhibit diverse reactions to temperature patterns that influence their progression from the vegetative to the reproductive phase (Wanga *et al.* 2021).

Management of Soil Moisture

Alterations in soil moisture levels can bring about significant transformations in plant growth and developmental processes, impacting factors like plant height, time to flowering, seed production, and maturation (Anjum *et al.* 2017; Hussain *et al.* 2018). Instances of drought or flooding stress can induce an expedited flowering and maturation cycle, a phenomenon applicable in speed breeding techniques. Among these, drought stress is the most frequently employed method for crops such as wheat, barley, and pearl millet (Shavrukov *et al.* 2017). In the case of pearl millet, drought stress leads to an early flowering stage, potentially evolved as an adaptive "escape mechanism" to ensure the production of the succeeding generation (Vadez *et al.* 2012). These investigations underscore the potential to enhance water supply management within crop speed breeding facilities, aiming to achieve a more efficient turnover of generations.

Carbon Dioxide (>400 ppm)

Elevated levels of carbon dioxide (CO_2) have the potential to accelerate plant growth and expedite the transition from the vegetative to the reproductive phase in angiosperms (Jagadish *et al.* 2016). However, distinct crop species exhibit varying reactions to elevated CO_2 concentrations. While increased CO_2 levels had minimal impact on flowering timing, they did augment flower production, which proves advantageous for generating a larger number of crosses. Manipulating CO_2 levels necessitates suitable infrastructure like growth chambers, CO_2 cylinders, regulators, and entails operational expenses. Furthermore, strict adherence to health protocols and safety guidelines is imperative when handling and employing CO_2 cylinders and valves (Wanga *et al.* 2021).

One of the most consistent outcomes of enhanced atmospheric CO_2 on plants is an amplified rate of photosynthetic carbon assimilation by leaves. Plant growth under elevated CO_2 concentrations ranging from 475 to 600 ppm yields an average increase of 40% in leaf photosynthetic rates. CO_2 levels also play a crucial role in modulating the aperture of stomatal pores through which plants exchange gases with the external environment. Nonetheless, the effects of heightened CO_2 are not uniform; certain species, particularly those employing the C4 variant of photosynthesis, exhibit a lesser response to elevated CO_2 due to a decrease in stomatal conductance. This adjustment may indirectly enhance photosynthesis by helping plants avert water stress during drought conditions.

Plant Nutrition and Hormones

Plant nutrition and hormones have been used to accelerate growth and induce flowering and seed set, and germination of immature seed in vitro (Bermejo *et al.* 2016). Varied responses to plant growth regulators (PGRs) are achieved when used in controlled environments such as greenhouses and growth chambers in which the photoperiod and temperatures can be monitored and controlled. Auxins, gibberellins and cytokinins plays a pivotal role. When cytokinin levels are lower than auxin levels, the plant is in vegetative growth. As cytokinin levels increase and auxin levels decrease, the plant transitions into the reproductive growth stage. Gibberellins help to control the transition from vegetative to reproductive growth. Nitrogen results in increased growth and higher biomass yields (Wanga *et al.* 2021).

Density of Plant Population

High-density planting involves cultivating crops at higher densities than what is needed to achieve maximum yield. This intensified planting arrangement leads to increased plant height as a consequence of heightened competition for light. As a result, there is an accelerated shift from the vegetative phase to the reproductive growth stages (Warnasooriya and Brutnell, 2014). This strategy proves advantageous in triggering early flowering and maturation, consequently augmenting the frequency of generation cycles within a single year. It is essential to conduct initial trials to determine the specific high-density planting conditions suitable for a given genotype, aiming to effectively promote early flowering in the context of speed breeding. High-density planting represents an economical speed breeding approach that facilitates swift generational progress while upholding the substantial population size necessary for advanced selections (Wanga *et al.* 2021).

Table 1 lists examples of some of the crops developed by speed breeding.

Table 1: Some crops developed by speed breeding

Crop	**Type of photoperiod**	**Techniques**	**Days to flowering**	**Selection method**	**No. of generation/ year**	**Reference**
Rice	Short day plant	Photoperiod, temperature and high-density planting	75–85	Single seed descent (SSD)	4–5	Rana *et al.* 2019, Collard *et al.* 2017

Barley	Long day plant	Photoperiod, temperature, soil fertility, immature seed germination and embryo rescue	24–36	SSD	9	Zheng *et al.* 2013
Wheat	Long day plant	Photoperiod, temperature, soil fertility, immature seed germination and embryo rescue	28–41	SSD	7.6	Zheng *et al.* 2013
Amaranth	Short day Plant	Photoperiod and temperature	28	SSD	6	Stetter *et al.* 2016
Faba bean	Long day Plant	Plant hormones, light intensity, photoperiod and immature seed	29–32	Single pod descent (SPD)	7	Mobini *et al.* 2015
Pea	Long day Plant	Plant hormones, photoperiod and immature seed germination	33	–	5	Mobini and Warkentin 2016
Groundnut	–	Temperature and photoperiod	25–27	SPD	3	O'Connor *et al.* 2013
Chick pea	Long day plant	Photoperiod and immature seed germination	33	SPD	7	Samineni *et al.* 2019
Soya bean	Short day plant	Photoperiod, temperature and immature seed germination	23	SSD	5	Jahne *et al.* 2020
Brassica oleracea	Long day plant	22-hour photoperiod	108	–	–	Ghosh *et al.* 2018
Brassica napus	Long day plant	22-hour photoperiod	87	–	–	Ghosh *et al.* 2018
Brassica rapa	Long day plant	22-hour photoperiod	87	–	–	Ghosh *et al.* 2018
Chili	Day neutral plant	Photoperiod, Light intensity	38–40	–	4	Liu *et al.* 2022

Importance/Advantages of Speed Breeding

Multiple Generations in 1 Year

Speed breeding, a cutting-edge plant cultivation technique, offers compelling reasons for achieving multiple generations in a single year. Primarily, this approach leverages carefully controlled growth conditions, such as extended light exposure through artificial lighting systems. By mimicking longer daylight hours, plants receive signals that accelerate their growth and reproductive processes, enabling them to progress through their life cycles at an expedited pace. Additionally, controlled temperature and humidity levels optimize plant metabolic activity and promote rapid development. This not only enables quicker seed germination but also accelerates flowering and seed set. Moreover, speed breeding reduces the time required for selecting desirable traits, as multiple generations can be observed and evaluated within a shorter timeframe. This shortened breeding cycle facilitates the identification and fixation of beneficial traits, contributing to the rapid development of improved plant varieties (Fig. 1). By addressing the challenge of lengthy generation times in conventional breeding, speed breeding holds immense promise for advancing crop research, enhancing genetic gains, and ultimately addressing global food security concerns (Collard *et al.* 2017, Watson *et al.* 2018, Saxena *et al.* 2019, Cazzolla *et al.* 2021).

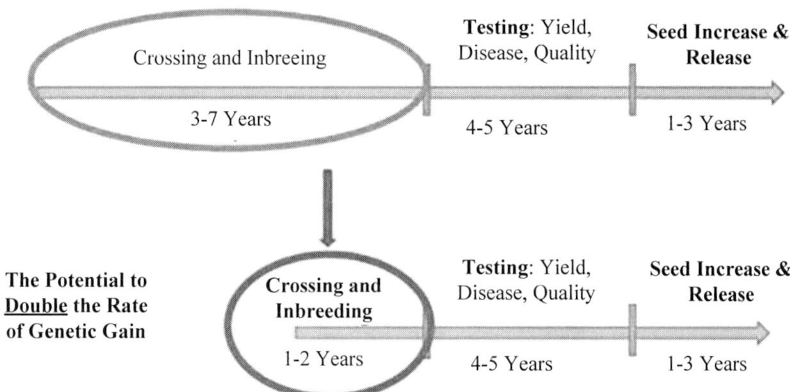

Fig. 1: Speeding up the breeding pipeline (Wanga *et al.*, 2021).

Fast Way to Obtain Fixed Homozygous Lines through Single Seed Descent Method

Speed breeding amenability with selection methods like single seed descent method, single plant selection method and single pod descent method (Fig.2).

The single seed descent (SSD) method, when combined with speed breeding, presents compelling advantages for swiftly obtaining fixed homozygous lines. SSD involves selecting a single seed from each generation and growing it separately, thus promoting the fixation of desired traits through self-pollination. In the context of speed breeding, these advantages are amplified. Rapid growth conditions, including extended light exposure and controlled environmental factors, accelerate plant life cycles. This means that within a short period, multiple generations can be progressed, enhancing the chances of homozygosity. The SSD method reduces the need for laborious and time-consuming steps such as manual pollination and crosses, streamlining the breeding process. Consequently, fixed homozygous lines can be achieved far more efficiently, enabling quicker trait fixation and expediting the development of improved plant varieties. This synergy between SSD and speed breeding not only saves time and resources but also contributes to the advancement of agricultural research and the timely creation of crops with enhanced attributes (Ghosh *et al.* 2018, Watson *et al.* 2018, Wanga *et al.* 2021).

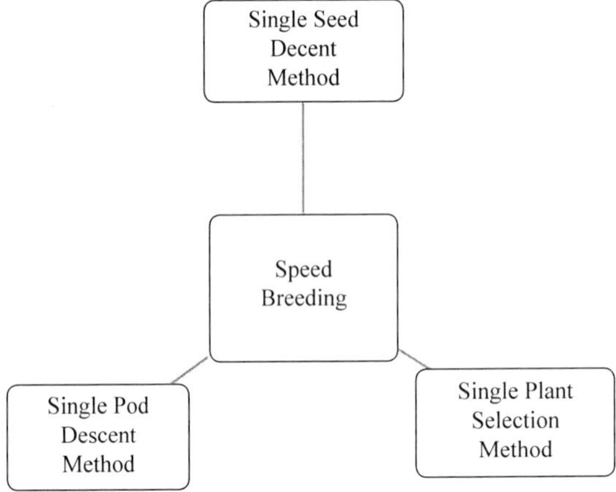

Fig. 2: Speed breeding amenability with selection methods (Wanga *et al.*, 2021).

Phenotypic Selection in Early Segregating Generations

Phenotypic selection within early segregating generations aligns seamlessly with the principles of speed breeding, offering distinct advantages. By subjecting plants to accelerated growth conditions and controlled environments, speed breeding enables the rapid advancement of generations. This expeditious lifecycle progression, combined with the implementation of phenotypic selection, allows for the swift identification and retention of desired traits. Early in the breeding process, visually evident characteristics can be evaluated and chosen, expediting the decision-making process. With the shortened breeding cycle, undesirable traits are efficiently eliminated, leading to the development of improved lines faster. Additionally, early phenotypic selection complements the fast-paced nature of speed breeding by ensuring that only the most promising candidates continue to subsequent generations. This synergistic approach accelerates the development of high-quality plant varieties with the desired attributes, contributing to enhanced agricultural productivity and genetic gains.

Rapid Introgression Genes into Elite Lines Using Marker-assisted Selection

Speed breeding coupled with many breeding methods but marker-assisted selection (MAS) finds a natural partner in speed breeding, offering compelling advantages for the swift introgression of genes into elite lines (Fig. 3). Speed breeding's controlled growth conditions accelerate plant generations, ensuring a quicker turnover. When coupled with MAS, the process gains precision and

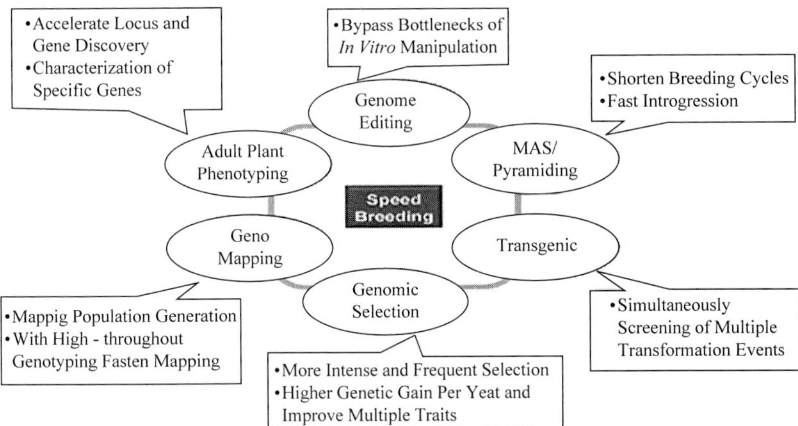

Fig. 3: Speed breeding coupled with other breeding methodologies Begna 2022. (MAS: maker-assisted selection)

efficiency (Fig. 3). Molecular markers enable the identification of specific genes associated with desirable traits, streamlining the selection process. This targeted approach eliminates the need for time-consuming phenotypic evaluations, allowing breeders to focus on the genetic aspects. Rapid generations facilitate the validation of marker–trait associations, ensuring accurate gene transfer in less time. The synergy between MAS and speed breeding accelerates the development of improved lines by swiftly transferring beneficial genes, reducing the breeding cycle. This expeditious introgression contributes to the timely creation of superior plant varieties, fostering enhanced agricultural productivity and resilience (Hickey *et al.* 2017, Wolter *et al.* 2019).

Study of Plant–Pathogen Interaction and Flowering Time

Speed breeding offers distinct advantages for studying plant–pathogen interactions and flowering time. The expedited growth conditions inherent to speed breeding enable researchers to rapidly observe the progression of plant responses to pathogens. This accelerated timeline allows for quicker identification of resistance or susceptibility traits, aiding in the development of disease-resistant varieties. Similarly, the controlled environment of speed breeding facilitates the manipulation of flowering time. By subjecting plants to extended light exposure, researchers can induce earlier or synchronized flowering, providing insights into flowering mechanisms and genetic factors controlling this crucial trait. The shortened growth cycle enhances the efficiency of these studies, enabling a deeper understanding of plant–pathogen dynamics and flowering time regulation within a condensed timeframe. This synergy between speed breeding and targeted research areas accelerates scientific discoveries and contributes to the advancement of crop improvement strategies.

Multi-environmental Trail Across Years

Speed breeding presents compelling advantages for conducting multi-environmental trials across multiple years. The rapid growth cycles within speed breeding enable the evaluation of plant performance across diverse environments in a shorter time frame. By simulating various conditions, such as temperature and light regimes, researchers can effectively mimic different growing seasons and geographical locations. This allows for a comprehensive assessment of a plant's adaptability and performance under varying circumstances, providing valuable insights into genotype–environment interactions. The efficient turnover of generations in speed breeding facilitates the accumulation of data over successive years, enabling the observation of trends and stability across environments. Additionally, the controlled conditions of speed breeding mitigate the uncontrollable variables that can confound field trials, leading to more reliable and reproducible results. The combination of speed breeding with multi-year, multi-environmental trials enhances the efficiency and accuracy of plant breeding efforts, aiding in the development of robust and adaptable crop varieties.

Integrated with Genomics Selection and Genome Editing

Integrating speed breeding with genomics selection and genome editing presents compelling advantages. Speed breeding's accelerated growth cycles align seamlessly with genomics selection, allowing for the rapid evaluation of genetic markers associated with desirable traits. By swiftly identifying and selecting individuals carrying these markers, the breeding process becomes more efficient, leading to the development of improved lines in less time. Similarly, the controlled conditions of speed breeding enhance the application of genome editing techniques. The expedited generations enable quicker validation of edited traits, optimizing the editing process. By accelerating the evaluation of edited plants, researchers can promptly fine-tune desired traits and expedite the development of precise genetic modifications. The combination of speed breeding, genomics selection, and genome editing creates a potent platform for enhancing the efficiency, accuracy, and speed of crop improvement efforts (Ahmar *et al.* 2020).

High-throughput Phenotypic Screens for Multiple Traits

Speed breeding offers a compelling platform for conducting high-throughput phenotypic screens for multiple traits. The rapid growth cycles facilitated by speed breeding expedite the assessment of diverse phenotypic characteristics within a shorter timeframe. This acceleration enables researchers to simultaneously evaluate a wide range of traits across numerous plant individuals or lines. By subjecting plants to controlled environments and extended light exposure, speed breeding enhances the visibility of traits, allowing for efficient and accurate measurements. The rapid turnover of generations in speed breeding enables the accumulation of substantial data in a relatively brief period, facilitating the identification of correlations and interactions among multiple traits. This streamlined approach not only accelerates the selection of desired traits but also allows for the discovery of unexpected associations. Through the integration of high-throughput phenotypic screens with speed breeding, researchers can efficiently unravel complex trait relationships and expedite the development of superior plant varieties tailored to specific needs (Alahmad *et al.* 2018).

Exploit Gene Bank Accessions and Mutant Collection for Rapid Gene Discovery

Speed breeding provides a powerful avenue for rapidly uncovering genes of interest by tapping into gene bank accessions and mutant collections. The accelerated growth cycles in speed breeding expedite the evaluation of a diverse array of plant materials. By subjecting gene bank accessions and mutants to controlled environments with extended light exposure, speed breeding enhances the manifestation of traits and enables efficient phenotypic screenings. This rapid assessment aids in the identification of novel genes associated with desired characteristics, leading to quicker gene discovery. Furthermore, speed breeding's quick generation turnover allows researchers to rapidly validate the function of identified genes and assess their potential contributions to plant performance. The synergy between speed breeding and gene bank resources expedites the process of identifying and leveraging valuable genetic assets, accelerating the development of improved plant varieties with enhanced attributes.

Challenges of Speed Breeding

Lack of Trained Plant Breeders and Breeding Technicians

A key hurdle to the widespread adoption of speed breeding in the public sector arises from a shortage of trained plant breeders and technicians in developing nations (Morris *et al.* 2006; Shimelis *et al.* 2019). This issue is compounded by a high turnover of personnel from public breeding programs to private seed companies due to better compensation. Additionally, specialized plant breeding expertise is limited due to the scarcity of postgraduate programs in developing countries. In some places, the absence of appropriate legislative and administrative frameworks hampers plant breeders' rights and seed regulation, discouraging progress along the value chain (Tripp *et al.* 2007). To address this, developing nations must revise their policies concerning investment in plant breeding education, research, and staff retention to ensure the success of long-term crop enhancement efforts and the integration of innovations like speed breeding.

Inadequate Infrastructure

Speed breeding relies on advanced infrastructure to control factors like soil moisture, temperature, and photoperiod. However, limited institutional support in public breeding programs in developing countries hampers the adoption of innovative methods like speed breeding and biotechnology. Additionally, scarcity of specialized equipment for early trait selection poses a challenge. Overdependence on donor agencies and the absence of coordination among regional programs lead to duplication and inefficient resource use. To address this, collaboration between national and regional organizations is crucial for infrastructure development and knowledge sharing. A cost-effective approach could involve repurposing shipping containers with solar-powered controls for temperature and light (Watson *et al.* 2018, Wanga *et al.* 2021, Samantara *et al.* 2022).

Unreliable Water and Electricity Supplies for Sustainable Operations

In public plant breeding programs, managing temperature and photoperiod for speed breeding is hindered by unreliable electricity supplies (O'Connor *et al.* 2013).

Lack of Standard Protocols

The absence of standardized protocols in speed breeding can be attributed to the intricate interplay of multiple factors. One fundamental factor is the inherent diversity among plant species and their respective varieties, each possessing distinct growth requirements. Consequently, creating a uniform set of protocols that adequately addresses the nuanced necessities of diverse crops becomes inherently complex. Moreover, the evolving nature of speed breeding as a relatively novel technique contributes to the lack of universally accepted protocols. Ongoing research seeks to refine the process for different plant types, resulting in a dynamic landscape of methodologies (Wanga *et al.* 2021, Cazzolla *et al.* 2021, Kamenya *et al.* 2021).

High Initial Investment

The high initial investment in speed breeding is met with challenges due to its substantial costs that can strain budgets and resources. The adoption of speed breeding demands significant financial outlays for specialized equipment, controlled environments, and advanced technologies, potentially posing a barrier to entry for smaller or resource-constrained agricultural entities. Moreover, the complex infrastructure required for speed breeding might necessitate additional training and expertise, leading to increased operational costs (Ghosh *et al.* 2018).

Early Harvest of Immature Seeds Interferes with Phenotyping of Different Seed Traits

Early harvesting of immature seeds can significantly disrupt the accurate phenotyping of various seed traits due to the incomplete development of crucial characteristics. Immature seeds often lack well-defined features, such as size, color, and texture, which are essential for precise trait evaluation. Premature harvesting can lead to inaccurate measurements of parameters, such as seed weight, shape, and coat properties, undermining the reliability of phenotypic data. Moreover, crucial physiological and biochemical processes, such as nutrient accumulation and metabolite composition, might not have occurred fully in immature seeds, distorting the representation of seed quality and composition. Therefore, allowing seeds to reach their natural maturity before harvest is crucial for obtaining comprehensive and trustworthy phenotypic information, ensuring the validity of subsequent research and breeding decisions (Ghosh *et al.* 2018, Watson *et al.* 2018, Samantara *et al.* 2022).

Different Responses of Different Plant Species When Exposed to Extended Photoperiod

The varying responses of different plant species to extended photoperiods in speed breeding can be attributed to their distinct physiological and genetic characteristics. Plant species have evolved under diverse environmental conditions, resulting in unique photoperiod sensitivities that influence their growth and development. Extended photoperiods in speed breeding can disrupt the natural circadian rhythms and developmental processes of some species, potentially causing physiological stress and reduced growth rates. Additionally, genetic factors such as the presence or absence of specific flowering genes can dictate a plant's responsiveness to changes in day length, leading to inconsistent flowering and fruiting responses among different species. Furthermore, variations in the optimal light intensity and spectral composition required for growth can influence how different plants utilize extended photoperiods, further contributing to differential outcomes. Therefore, understanding and accommodating the intrinsic differences in plant species genetic makeup and physiological mechanisms is essential to effectively harness the benefits of extended photoperiods in speed breeding and ensure successful and uniform crop development across a range of species (Ghosh *et al.* 2018, Osnato *et al.* 2022).

Disease Outbreak in Controlled Environment Conditions

Disease outbreaks in controlled environment conditions during speed breeding can arise due to several factors. The tightly controlled and optimized growth conditions in speed breeding can inadvertently create an environment conducive to the rapid proliferation of certain pathogens. Reduced air circulation and high humidity levels, often necessary to maximize plant growth rates, can create a microclimate favorable for fungal and bacterial pathogens to thrive. Moreover, the close proximity of plants in controlled environments can facilitate the easy spread of diseases through direct contact or airborne transmission. The accelerated growth cycles in speed breeding might not allow for sufficient time for the development of robust immune responses in plants, rendering them more susceptible to infections. Furthermore, the reliance on a limited number of genetically uniform plants in speed breeding can amplify the impact of a disease outbreak, potentially leading to significant crop losses. Therefore, implementing strict hygiene practices, incorporating disease-resistant varieties, and carefully monitoring plant health are essential strategies to mitigate the risk of disease outbreaks in controlled environment speed breeding setups (Samantara *et al.* 2022; Sharma *et al.* 2023).

Incorporation of Relatively Inherited Traits

Incorporating relatively inherited traits in speed breeding, while seemingly advantageous, can have drawbacks that need consideration. Relying solely on traits with higher heritability might lead to a narrower focus on a limited set of characteristics, potentially overlooking valuable but less heritable traits that contribute to overall plant performance and adaptability. This approach could inadvertently result in reduced genetic diversity within the breeding population, making the crop more susceptible to diseases, pests, and changing environmental conditions. Furthermore, prioritizing relatively inherited traits might lead to unintended trade-offs, where the improvement of one trait comes at the expense of other crucial agronomic or quality traits. Additionally, if genetic markers associated with these traits are not fully validated, there is a risk of introducing unintended genetic variations that could have negative consequences. Balancing the incorporation of relatively inherited traits with broader trait considerations is essential to ensure a well-rounded and resilient crop that meets diverse agricultural and consumer needs (Hickey *et al.* 2017).

Future Outlook

While modern plant breeding currently relies on established techniques, the introduction of novel methods is poised to significantly enhance its efficiency and effectiveness. Looking ahead, a diverse array of approaches is anticipated to be developed, drawing from interdisciplinary principles to amplify their advantages. These encompass strategies for crop production, breeding methodologies, field testing approaches, genotyping technologies, and even the refinement of equipment and facilities. The aim is to apply these advancements across various crop species, ensuring the preservation of a varied food, fiber, and biobased economy. Speed breeding combined with SSD is more frequently utilized to create excellent inbred lines and efficient breeding lines that can be used to generate superior crop varieties faster and more cheaply than di-haploid production (Alahmad *et al.* 2018). The concept of "speed breeding" or rapid generation advancement (RGA) will play a pivotal role in expediting the integration of technological progress into future crop varieties (Ghosh *et al.* 2018; Watson *et al.* 2018). Additionally, the precision breeding capabilities brought about by gene editing offer a pathway for enhancing desirable traits with utmost accuracy (Liang *et al.* 2017). The potential for multiplex editing, a technology akin to that employed in the development of GMO (genetically modified organisms) crops, holds promise for further shortening the breeding cycle in conjunction with speed breeding and tissue culture-free techniques. This innovative breeding approach goes beyond merely accelerating generations faster than conventional field or greenhouse conditions. It also addresses a crucial challenge in plant breeding research, which is the time-intensive nature of the process. Its potential can be fully harnessed by harmonizing it with other molecular tools in agricultural research, such as CRISPR (clustered regularly interspaced palindromic repeats) and genomic selection. Recently, several open-access technologies have emerged, enabling plant breeders to utilize this method for their research. The central issue in plant research programs lies in generation time, where speed breeding emerges as a viable and practical solution. It enables crops to swiftly adapt to emerging threats like new diseases and changing environments. According to the originator of the speed breeding concept, its potential is virtually limitless. Ongoing efforts are exploring ways to integrate speed breeding with other contemporary crop breeding techniques. Despite its considerable promise, the speed breeding system necessitates well-equipped infrastructure and specialized tools, which could pose a challenge in the widespread adoption of this forward-looking concept (Sharma 2019). The emergence of transformative technologies like CRISPR/Cas9, CRISPR/Cpf1, base editing, and RGA has sparked a revolution in molecular biology, ushering in innovative applications within agriculture and marking a pivotal moment in plant breeding and cultivation.

Collectively, CRISPR-based gene drive systems are poised to eventually bring about advantages for humanity. These systems hold the potential to curb epidemics, enhance agricultural techniques, and manage the proliferation of invasive species. As part of these advancements, genetically modified plant varieties that resist insect pests, pathogens, and herbicides can be cultivated. By incorporating techniques like speed breeding, the efficacy of current genome editing methods can be elevated. For instance, CRISPR/Cas9 can target genes responsible for delayed flowering, expediting the process. Following the successful integration of Cas9 into plants, the resulting transgenic plants can be cultivated under accelerated conditions, bypassing traditional greenhouse settings and enabling the production of transgenic seeds within a shorter timeframe sometimes under a year. This approach not only facilitates the generation of stable and consistent homozygous traits, but also accelerates the breeding cycle, a departure from the multi-year timeline usually associated with GMO crop development.

Moreover, the synergy of these technologies could usher in a revolutionary increase in the pace of genetic improvement by implementing more efficient breeding strategies. Consequently, CRISPR/Cas9, centered on genome editing and expedited breeding techniques, is likely to witness a surge in popularity. It is poised to play a pivotal role in the creation of plants harboring specific desirable characteristics, thereby contributing to the global aim of eradicating hunger. The trajectory of innovation often targets a select few economically significant crops, necessitating adaptations to their unique reproductive and propagation processes. Typically originating in developed countries and the private sector, the transfer of such technology should extend to the public domain and developing nations, given the substantial financial investment required for pioneering breakthroughs.

Conclusion

The plant research community is facing the challenge of meeting the demands of a growing global population and changing climate through enhanced plant improvement efforts. To address this, the integration of cutting-edge genomics methods with rapid gene fixation processes holds promise for significantly accelerating genetic advancements in breeding initiatives. One approach that has shown significant progress is speed breeding, which has expedited breeding programs for economically significant species. Through manipulation of light, temperature, and physical constraints, speed breeding expedites various stages of plant breeding, leading to quicker breeding cycles. Speed breeding facilitates the swift transition to homozygosity and efficient evaluation of existing or modified lines, such as gene-edited and transgenic crops. While speed breeding techniques streamline genotyping and phenotyping with reduced resource consumption, it is important to further investigate and address any adverse impacts of speed breeding conditions on plant growth and development. The established speed breeding protocols are adaptable for both small- and large-scale adoption, and they can be refined and tailored to local requirements and innovations. Speed breeding, combined with genetic tools and resources, enables plant biologists to scale up their research in the field of crop improvement. By progressively enhancing these protocols and integrating them with contemporary breeding methods, the potential to pinpoint and transfer crucial genes for crop resilience and adaptation can be fully realized. International collaborations involving interdisciplinary teams are crucial for promoting the incorporation of speed breeding systems into fundamental and applied research, especially in developing nations.

References

Ahmar, S., Gill R.A., Jung K.H., Faheem A., Qasim M.U., Mubeen M., et al. "Conventional and Molecular Techniques from Simple Breeding to Speed Breeding in Crop Plants: Recent Advances and Future Outlook." International Journal of Molecular Sciences 21, 7 (2020):2590.

Alahmad S., Dinglasan E., Leung K.M. "Speed Breeding for Multiple Quantitative Traits Indurum Wheat." Plant Methods 14 (2018):36.

Anjum, S.A., Ashraf U., Zohaib A., Tanveer M., Naeem M., Ali I., et al. "Growth and Developmental Responses of Crop Plants Under Drought Stress: A Review." Zemdirbyste 104, 3 (2017):267-76.

Anonymous 2022 IISD. World Population to Reach 9.9 Billion by 2050. Available Online: https://sdg.iisd.org/news/world-popoulation-to-reach-9.9-billion-by-2050/ (Accessed on August 2 2023).

Begna, T. "Speed Breeding to Accelerate Crop Improvement". International Journal of Agricicultural Science and Food Technology 8, 2 (2022):178-186.

Bermejo, C., Gatti, I., Cointry, E. "In vitro Embryo Culture to Shorten the Breeding Cycle in Lentil (Lens culinaris Medik)". Plant Cell, Tissue and Organ Culture 127, 3 (2016):585-90.

Cazzolla, F., Bermejo C.J., Guindon M.F., Cointry E. "Speed Breeding in Pea (Pisum sativum L.), An Efficient and Simple System to Accelerate Breeding Programs." Euphytica 216 (October 2020):178.

Collard, B.C.Y., Beredo J.C., Lenaerts B., Mendoza R., Santelices R., Lopena V., et al. "Revisiting Rice Breeding Methods–Evaluating the Use of Rapid Generation Advance (RGA) for Routine Rice Breeding." Plant Production Science 20, 4 (2017):337-52.

Ghosh, S., Watson A., Gonzalez-Navarro O.E., Ramirez-Gonzalez R.H., Yanes L., Mendoza-Suárez M., et al. "Speed Breeding in Growth Chambers and Glasshouses for Crop Breeding and Model Plant Research." Nature Protocols 13, 12 (2018):2944-63.

Hatfield, J.L., Prueger J.H. "Temperature extremes: Effect on Plant Growth and Development." Weather and Climate Extremes 10 (2015):4-10.

Hickey, L.T., German S.E., Pereyra S.A., Diaz J.E., Ziems L.A., Fowler R.A., et al. "Speed Breeding for Multiple Disease Resistance in Barley." Euphytica 213 (2017):64.

Hussain, H.A., Hussain S., Khaliq A., Ashraf U., Anjum S.A., Men S., et al. "Chilling and Drought Stresses in Crop Plants: Implications, Cross Talk, and Potential Management Opportunities." Frontiers in Plant Science 9 (2018):393.

Jagadish, S.V.K., Bahuguna R.N., Djanaguiraman M., Gamuyao R., Prasad P.V.V., Craufurd P.Q. "Implications of High Temperature and Elevated CO_2 on Flowering Time in Plants." Frontiers in Plant Science 7 (2016):913.

Jahne, F., Hahn V., Wurschum T., Leiser W.L. "Speed Breeding Short-day Crops by LED-controlled Light Schemes." Theoretical and Applied Genetics 133 (2020):2335-42.

Kamenya, S.N., Mikwa E.O., Song B., Odeny D.A. "Genetics and Breeding for Climate Change in Orphan Crops." Theoretical and Applied Genetics 134 (2021):1787-815.

Kouressy, M., Dingkuhn M., Vaksmann M., Heinemann A.B. "Adaptation to Diverse Semi-arid Environments of Sorghum Genotypes Having Different Plant Type and Sensitivity to Photoperiod." Agricultural and Forest Meteorology 148, 3 (2008):357-71.

Liang, Z., Chen K., Li T., Zhang Y., Wang Y., Zhao Q., et al. "Efficient DNA-free Genome Editing of Bread Wheat Using CRISPR/Cas9 Ribonucleoprotein Complexes." Nature Communications 8 (2017).

Liu, K., He R., He X., Tan J., Chen Y., Li Y., et al. "Speed Breeding Scheme of Hot Pepper through Light Environment Modification." Sustainability 14 (2022):122-5.

McClung, C.R., Lou P., Hermand V., Kim J.A. "The Importance of Ambient Temperature to Growth and the Induction of Flowering." Frontiers in Plant Science 7 (2016):1266.

Mobini, S.H., Lulsdorf M., Warkentin T.D., Vandenberg A. "Plant Growth Regulators Improve In Vitro Flowering and Rapid Generation Advancement in Lentil and Faba Bean." In Vitro Cellular and Developmental Biology - Plant 51, 1 (2015):71-9.

Mobini, S.H., Warkentin T.D. "A Simple and Efficient Method of In Vivo Rapid Generation Technology in Pea (Pisum sativum L.)." In Vitro Cellular and Developmental Biology - Plant 52, 5 (2016):530-6.

Morris, M., Edmeades G., Pehu E. "The Global Need for Plant Breeding Capacity: What Roles for the Public and Private Sectors?" HortScience 41, 1 (2006):30-9.

O'Connor, D.J., Wright G.C., Dieters M.J., George D.L., Hunter M.N., Tatnell J.R., et al. "Development and Application of Speed Breeding Technologies in A Commercial Peanut Breeding Program." Peanut Science 40, 2 (2013):107-14.

Osnato, M., Cota I., Nebhnani P., Cereijo U., Pelaz S. "Photoperiod Control of Plant Growth: Flowering Time Genes Beyond Flowering." Frontiers in Plant Science 12 (February 2022).

Rana, M.M., Takamatsu T., Baslam M., Kaneko K., Itoh K., Harada N., et al. "Salt Tolerance Improvement in Rice through Efficient SNP Marker-Assisted Selection Coupled with Speed-Breeding." International Journal of Molecular Sciences 20 (2019):2585.

Saito, H., Yuan Q., Okumoto Y., Doi K., Yoshimura A., Inoue H., et al. "Multiple Alleles at Early Flowering 1 Locus Making Variation in the Basic Vegetative Growth Period in Rice (Oryza sativa L.)." Theoretical and Applied Genetics 119, 2 (2009):315-23.

Samantara, K., Bohra A., Mohapatra S.R., Prihatini R., Asibe F., Singh L., et al. "Breeding More Crops in Less Time: A Perspective on Speed Breeding." Biology 11 (2022):275.

Samineni, S., Sen M., Sajja S.B., Gaur P.M. "Rapid Generation Advance (RGA) in Chickpea to Produce Up to Seven Generations Per Year and Enable Speed Breeding." Crop Journal 8, 1 (2019):164-9.

Saxena, K.B., Saxena R.K., Hickey L.T., Varshney R.K. "Can A Speed Breeding Approach Accelerate Genetic Gain in Pigeonpea?" Euphytica 215, 12 (2019):1-7.

Sharma, A., Jones J.B., White F.F. "Recent Advances in Developing Disease Resistance in Plants." F1000Research 8, (2019):1934.

Sharma, S., Kumar A., Dhakte P., Raturi G., Vishwakarma G., Barbadikar K.M., et al. "Speed Breeding Opportunities and Challenges for Crop Improvement." Journal of Plant Growth Regulation 42 (2023):46-59.

Shavrukov, Y., Kurishbayev A., Jatayev S., Shvidchenko V., Zotova L., Koekemoer F., et al. "Early Flowering As A Drought Escape Mechanism in Plants: How Can It Aid Wheat Production?" Frontiers in Plant Science 8 (2017):1950.

Shimelis, H., Gwata, E. T., Laing, M. D. (2019) "Crop Improvement for Agricultural Transformation in Southern Africa". In R. A. Sikora, E. R. Terry, P. L. G. Vlek, & J. Chitja (Eds.), Transforming Agriculture in Southern Africa, 1st, ed. (pp. 97-103). Routledge.

Stetter, M.G., Zeitler L., Steinhaus A., Kroener K., Biljecki M., Schmid K.J. "Crossing Methods and Cultivation Conditions for Rapid Production of Segregating Populations in Three Grain Amaranth Species." Frontiers in Plant Science 7 (June 2016).

Tripp, R., Louwaars N., Eaton D. "Plant Variety Protection in Developing Countries. A Report from the Field." Food Policy 32, 3 (2007):354-71.

Vadez, V., Hash T., Bidinger F.R., Kholova J. "Phenotyping Pearl Millet for Adaptation to Drought." Frontiers in Physiology 3 (2012):386.

Vince-Prue, D. 1994. The Duration of Light and Photoperiodic Responses. Photomorphogenesis in Plants. Edited by Kendrick R.E., Kronenberg G.H.M. Dordrecht: Springer.

Wanga, M.A., Shimelis H., Mashilo J., Laing M.D. "Opportunities and Challenges of Speed Breeding: A Review." Plant Breeding 140 (March 2021):185-94.

Warnasooriya, S. N., Brutnell T. P. "Enhancing the Productivity of Grasses under High-density Planting by Engineering Light Responses: From Model Systems to Feedstocks". Journal of Experimental Botany 65, 11 (2014), 2825-34.

Watson, A., Ghosh S., Williams M.J., Cuddy W.S., Simmonds J., Rey M.D., et al. "Speed Breeding is A Powerful Tool to Accelerate Crop Research and Breeding." Nature Plants 4 (2018):23-9.

Wolter, F., Schindele P., Puchta H. "Plant Breeding at the Speed of Light: The Power of CRISPR/Cas to Generate Directed Genetic Diversity at Multiple Sites." BMC Plant Biology 19 (2019):1-8.

Zheng, Z., Wang H.B., Chen G.D., Yan G.J., Liu C.J. "A Procedure Allowing Up to Eight Generations of Wheat and Nine Generations of Barley Per Annum." Euphytica 191, 2 (2013):311-16.

15

The Role of Epigenetics in Plant Breeding

***Sahil G. Shamkuwar*[1], *Akshay R. Uike*[2], *Dharm Veer Singh*[3], and Kamaluddin*[4]**

[1]*Department of Genetics and Plant Breeding, Institute of Agricultural Sciences Banaras Hindu University, Varanasi, Uttar Pradesh, India*

[2]*Department of Plant Breeding and Genetics, Assam Agricultural University, Jorhat, Assam, India*

[3,4]*Department Genetics and Plant Breeding, Banda University of Agriculture and Technology, Banda, Uttar Pradesh, India*

Abstract

Epigenetics is a new field of research in plant biology, which aims to provide a powerful tool to improve crop productivity, adaptability, and resistance to biotic and abiotic stresses. In recent years, researchers have begun to explore the potential of epigenomics in plant breeding, providing breeders with an alternative to genetic engineering, addressing concerns related to genetically modified organisms (GMOs) in agriculture. In particular, epigenetic markers (epimarkers) are chemicals which regulate genes by sitting over the DNA sequence; they explore the consequences of thermal and ionic variation on the behavior of plant under stress conditions. Moreover, epialleles may be promising sources of new variation of traits that are controlled by genes that are not detrimentally regulated. In this chapter, we provide an overview of the current state of the art in epigenomic research, focusing on the impact of next-generation sequencing (NGS) on epigenomes, as well as the potential use of epigenetics in crop breeding.

Keywords: *Epigenetics, epialleles, epigenomics, breeding*

Introduction

Epigenetics is the study of heritable changes in gene expression that occur without alterations in the DNA sequence. It involves modifications of the DNA or associated proteins that can influence gene activity and expression. Most cells in an organism share the same genes, although they can be very diverse from one another. This is due to the fact that not all genes are active in all cells at the same time. The epigenetic context of a gene, or its accessibility to transcription, can tell whether the gene is likely to be active or not (Klemm *et al.* 2019). Indeed, the nuclear genome is found as a dense combination of long DNA strings wrapped around specialized proteins called histones to create nucleosomes (Samo *et al.* 2021). These nucleosomes can be further packed together to form even denser 3D formations that are sometimes visible under a microscope, giving rise to the term "chromatin". The modifications and proteins associated with the DNA are called epigenetic modifications, or marks, and they can influence gene expression within the cell. Epigenetic modifications can occur spontaneously in the genome and can be lost just as randomly presenting the concept of stability. As a result, the effects of epigenetic alterations on gene expression can only be transferred from mother cell to daughter cell and from generation to generation if the underlying modifications are actively maintained. To induce variability at the phenotypic level, both genome sequence dependent (genetic) and independent (epigenetic) variability are exploited in combination as an adaptation method (Tirnaz and Batley 2019) This, in turn, maximizes the chances of survival of at least some progeny under different conditions and provides a wider pool of phenotypic variation to be used for plant breeding (Gallusci *et al.* 2017).

Epigenetic modifications play a crucial role in various biological processes, including development, differentiation, and response to environmental stimuli. In recent years, researchers have begun to explore the potential of epigenetics in plant breeding to improve crop productivity, adaptability, and resistance to biotic and abiotic stresses. The main benefit derived from the dynamic nature of epigenetic networks is associated with the stability and reversibility of chromatin modifications. The stability of these changes is essential for maintaining epigenetic memory (somatic memory and inter/transgenerational memory), which allows cells to maintain their identity during plant development and "remember" favorable alterations leading to a selective advantage. The reversible nature of epigenetic changes confers plasticity, enabling differentiated cells to regain totipotency, and adequately respond and adapt to internal and environmental stimuli (Singh *et al.* 2019).

Epigenetics: History and Evolution

The field of epigenetics has a rich history dating back to the early 20th century. The hypothesis of epigenetic changes affecting the expression of chromosomes was put forth by the Russian biologist, Nikolai Koltsov (Morange *et al.* 2011). From the generic meaning, and the associated adjective epigenetic, British embryologist, C.H. Waddington coined the term epigenetics in 1942 as pertaining to epigenesis, in parallel to Valentin Haecker's "phenogenetics" (Waddington 1942), to describe "the interaction of genes with their environment, which bring the phenotype into being" (Waddington 1940). Two years later, Conrad Waddington pointed out that "It is possible that an adaptive response can be fixed without waiting for the occurrence of a mutation". Thus, epigenetic modifications are heritable and reversible modifications that significantly affect gene expression without any change in the nucleotide sequence of DNA (Meccariello 2019).

Epigenesis in the context of the biology of that period referred to the differentiation of cells from their initial totipotent state during embryonic development. When Waddington coined the term, the physical nature of genes and their role in heredity was not known. He used it instead as a conceptual model of how genetic components might interact with their surroundings to produce a phenotype; he used the phrase "epigenetic landscape" as a metaphor for biological development. Waddington held that cell fates were established during development in a process he called "canalization" much as a marble rolls down to the point of lowest local elevation (Waddington (2014). Waddington suggested visualizing the increasing irreversibility of cell type differentiation as ridges rising between the valleys where the marbles (analogous to cells) are travelling (Hall 2004).

Since then, significant advancements have been made in understanding the mechanisms underlying epigenetic modifications, such as DNA methylation, histone modifications, and noncoding RNA molecules. These modifications can be stable and heritable, leading to transgenerational effects on gene expression and phenotype. In recent times, Waddington's notion of the epigenetic landscape has been rigorously formalized in the context of the systems dynamics state approach to the study of cell fate. Cell fate determination is predicted to exhibit certain dynamics, such as attractor–convergence (the attractor can be an equilibrium point, limit cycle, or strange attractor) or oscillatory. (Rabajante and Babierra 2015).

Thus, epigenetics is a fascinating field of genetics, completely meddling with classical knowledge of the interaction between the genotype and the phenotype, and puzzling scientists for decades. According to the definition of Arthur Riggs and his colleagues, "epigenetics is the study of mitotically and/or meiotically heritable changes in gene function that cannot be explained by changes in DNA sequence" (Feil 2008). Epigenetic changes, which involve DNA methylation, histone modifications, chromatin remodeling, and activity of small RNAs (sRNAs), are thus heritable, but do not follow the known patterns of inheritance (Bräutigam *et al.* 2013).

Methods and Mechanism of Epigenetic Modification

Methods to Modify the Plant Epigenome

The methods used so far in the laboratory to modify the plant epigenome in order to obtain novel phenotypes are based on the use of chemicals, biotic and abiotic stresses, tissue culture, mutants, and grafting, or are molecular RNA-based methods, such as RNA interference (RNAi) and CRISPR (clustered regularly interspaced short palindromic repeats) (Flowchart 1).

Flowchart 1: Methods for epigenome modification in plants. (ABA: abscisic acid; CRISPR: clustered regularly interspaced short palindromic repeats; dCas9: dead Cas9; Exo-RNAi: exogenous RNA interference; HDAC: histone deacetylase; Int-hpRNA: intronic hairpin RNA; JA: ??; Nuc-hpRNA: nuclear hpRNA; Pol II: polymerase II; ROS: reactive oxygen species; TE: transposable element)

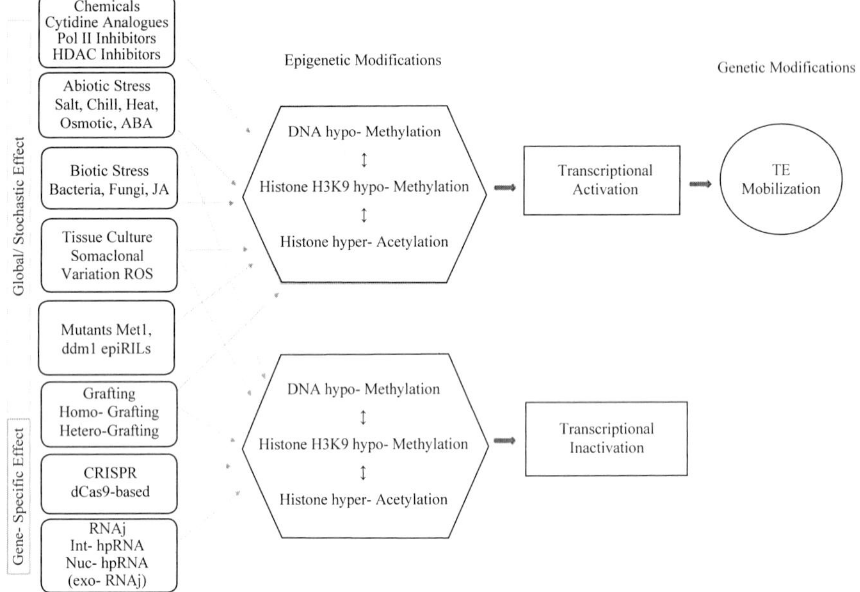

The plant epigenome may be modified using chemicals, abiotic and biotic stress, tissue culture, grafting, mutants, and CRISPR- and RNAi-based methods. Among these, only CRISPR and RNAi lead to gene-specific epigenetic modifications, while the exogenous application of nuc-hpRNAs (nuclear hairpin RNAs) is a completely transgene-free technology.

Notably, in the cases of DNA hypomethylation (tightly related to histone hypomethylation/hyperacetylation), global transcriptional activation will awaken silenced transposable elements (TEs), whose mobilization and translocation in the genome will ensure heritable genetic changes and additional phenotypic variability.

Mechanisms of Epigenetic Modification

Epigenetic modifications can occur through various mechanisms, including DNA methylation, histone modifications, and RNA-based processes. DNA methylation involves the addition of a methyl group to the DNA molecule, typically at cytosine residues. This modification can lead to gene silencing or activation, depending on its location within the genome. Histone modifications, on the other hand, involve the addition or removal of chemical groups to histone proteins, which package DNA into a compact structure. These modifications can alter the accessibility of genes for transcription factors and other regulatory proteins. Additionally, non-coding RNA molecules, such as small interfering RNAs (siRNAs) and microRNAs (miRNAs), can also influence gene expression by targeting specific mRNA molecules for degradation or translational repression.

Epigenetics mechanism includes:

- DNA methylation
- Histone modifications (acetylation, methylation, phosphorylation, ubiquitination, biotinylation, and simulation)

DNA methylation: DNA methylation targets the promoter region of the gene. When methyl group is added to DNA it stops the gene from being "seen" by forming 5-methylcytosine (5-mC). The methyl groups basically block other proteins from binding to the promoter region. These regions contain a high number of C (cytosine) and G (guanine) DNA bases, the cytosine or C residues in these regions can receive methyl groups added to them, then the gene sequence cannot be read so genes are switched off and no transcription after methylation (Moore *et al.* 2012). Changes in DNA methylation in response to drought stress cause differential expression of stress-responsive genes in the drought-tolerant variety of rice (Wang *et al.* 2004). DNA methylation is one of the most well-known epigenetic marks, it can be found in the coding regions of genes in many organisms, ranging from plants to humans, and it targets the promoter region of the genes; this process is catalyzed by an enzyme known as DNA methyltransferase (DNMT).

Histone modifications: Histone modification is another epigenetic modification. Histone octamers consist of two of each of the core histones H2A, H2B, H3, and H4, around which wraps ~146 bp of DNA, forming a nucleosome that is a basic unit of chromatin. Chromatin structure dictates transcriptional activity and inactivity, and post translational histone modifications including methylation, acetylation, phosphorylation, and ubiquitination of the N-terminal tails of the core histones are involved in alteration of chromatin structure. Generally, adding an acetyl group to a lysine residue of histones by histone acetyltransferase (HAT) is associated with transcriptional activation, while removal of the acetyl group by histone deacetylase (HDAC) is associated with transcriptional expression. In the case of histone methylation, lysine residues are methylated and the number of methyl groups (e.g., mono-, di-, or tri-methylation) is typically associated with transcriptional activation or repression. These changes are catalyzed by the histone lysine methyltransferase (HKMTase) and histone demethylases (HDMases). For example, trimethylation of lysine 4 of histone H3 (H3K4me3) and H3K36me3 are often associated with transcriptional activation, while H3K9me2 and H3K27me3 are associated with transcriptional repression (He *et al.* 2011; Quadrana and Colot 2016; Talbert and Henikoff 2021).

Need of Epigenetics in Plant Breeding?

Epigenetic modifications have been shown to play a crucial role in plant responses to environmental cues, including temperature, light, nutrients, and pathogens. By understanding and manipulating these epigenetic marks, plant breeders can potentially enhance crop performance under adverse conditions. For example, researchers have found that altering DNA methylation patterns can improve drought tolerance in crops like maize and rice. Similarly, modifying histone acetylation levels can enhance resistance to fungal diseases in wheat and barley. These findings highlight the potential of epigenetic approaches in plant breeding.

Classical Breeding to Epibreeding

Classical crop breeding refers to the transfer of desirable traits to cultivated crops by crossing and recurrent selection of genetic variants. Natural and induced genetic diversity has been the driving force behind it. However, the reliance on limited germplasm has resulted in an irreversible loss of genetic diversity, known as genetic erosion (Gallusci *et al.* 2017). This alarming narrowing of the genetic base has rendered crop genetic improvement difficult, exposing the vulnerability of current varieties to rapid climate changes and threatening world food security. Yet, traits are governed not only by genetics but by epigenetics as well. Thus, in view of the ever-growing genetic erosion, exploitation of epigenetic variations and/or the manipulation of the epigenome may constitute a promising breeding strategy to improve adaptation to climate change and ensure crop yield and quality (Tirnaz and Batley 2019; Pecinka *et al.* 2020).

Arabidopsis thaliana, *Brachypodium distachyon*, *Lotus japonicus*, and *Medicago truncatula* are examples of species used as models in plant biology which have made significant contributions to our understanding to the mechanisms governing plant epigenetics. Other crop model species, such as *Oryza sativa*, *Zea mays*, *Triticum aestivum*, *Glycine max*, and *Solanum lycopersicum*, provides a rich source of knowledge regarding the regulators and phenomena of plant-specific epigenetics (Kakoulidou *et al.* 2021). The improvement of agriculturally desired developmental or stress-related crop traits will considerably benefit from the dissemination of fundamental information across several facets of plant responsiveness to environmental cues mediated by heritable epigenetic variation (Flowchart 2).

Flowchart 2: How the development of epigenetic data and tools will lead to epibred crops and new varieties in the field adapted to climate change. In blue are indicated data already supporting epigenetics for breeding, in brown the uses of epigenetics for crop improvement (under development), and in green the post-production steps until new varieties in the field (future challenge).

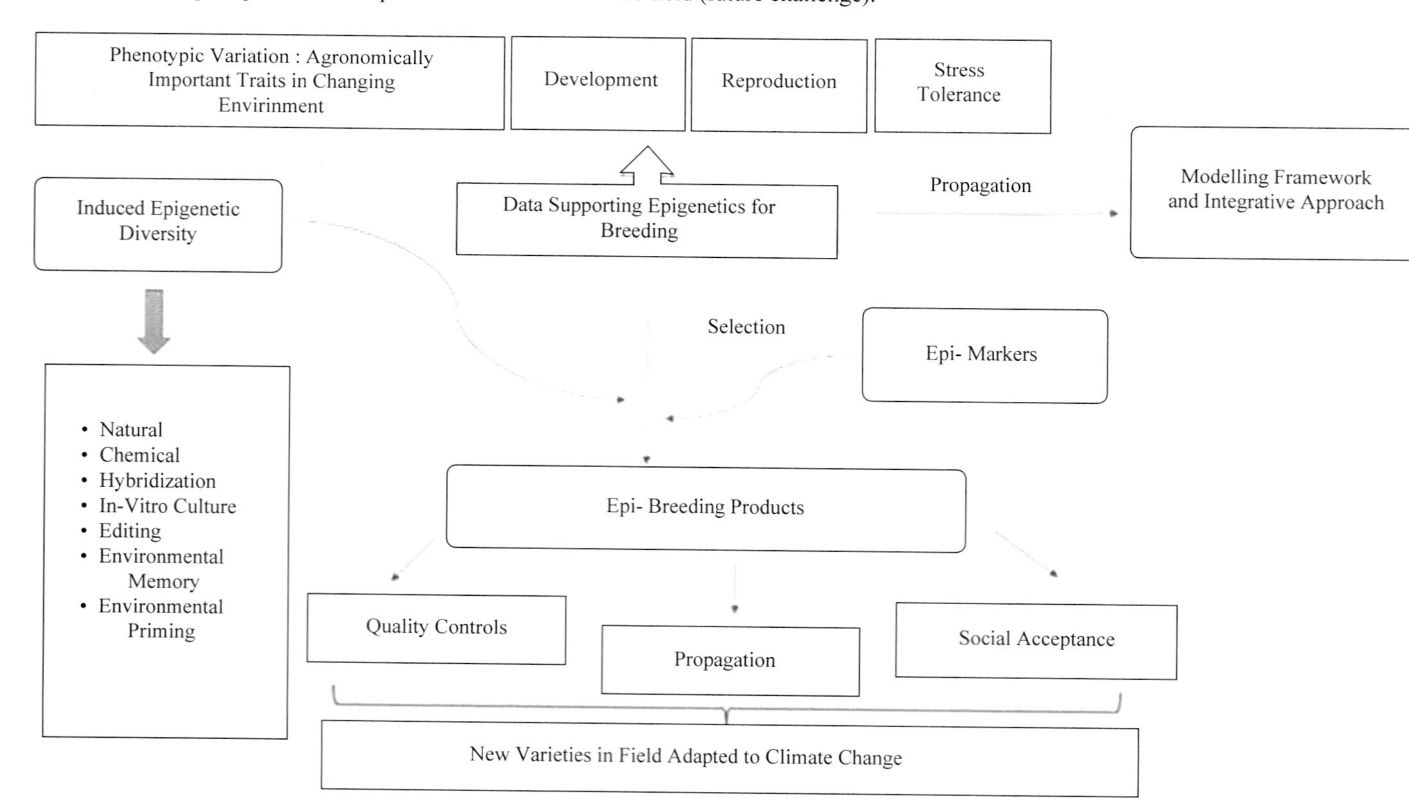

Drought tolerance in *Brassica napus*, bacterial resistance in *Arabidopsis*, blast disease resistance in *O. sativa*, and increased yield in *G. max* are all epigenetics-based and are just part of the growing body of evidence that underlines the high potential of epigenome exploitation for breeding purposes. On the one hand, analyzing the natural epigenetic diversity and unveiling its connection to various phenotypes could lead to the development of epigenetic markers (epimarkers) and the identification of epigenetic quantitative trait loci (epiQTL) associated to desired traits that could be employed for selection purposes at early stages in breeding programs. On the other hand, inducing artificial epigenetic diversity promises to deliver a great spectrum of phenotypic variability that can significantly invigorate breeding platforms (Varotto *et al.* 2020).

Natural Epigenetic Variation is Shaped by Developmental and Environmental Cues

The regulation of gene expression results from the combined action of reversible epigenetic mechanisms such as DNA methylation of cytosines, histone tail modifications, and noncoding small RNAs. It is a highly dynamic process that can be influenced by both internal (developmental, hormonal and nutritional) and external (biotic and abiotic) signals and which determines the progression of the lifecycle of the plant (Boyko *et al.* 2011; Finnegan *et al.* 2002). The study of environmentally induced epigenetic mechanisms has attracted a lot of attention since, by allowing quick adaptation to novel conditions through changes in gene expression (and thus, phenotypic plasticity), they could contribute to phenomena such as plant acclimation and stress response (Boyko *et al.* 2011). Vernalization is probably the most documented model in which epigenetic changes are triggered by a change in environmental conditions. In vernalization-responsive *Arabidopsis* accessions, flowering in spring conditions (long days and warm temperatures) depends on the inactivation of the flowering locus C (FLC) floral repressor, which itself is turned off through the time-dependent addition of repressive chromatin marks during exposure to winter-like conditions (short days and low temperatures) (Sung and Amasino 2004). In the annual *A. thaliana*, the epigenetic repression of FLC is highly stable throughout the life of the plant and is relieved between sexual generations so that the progeny of a vernalized plant requires vernalization for flowering. By contrast, in its perennial relative *Arabidopsis halleri*, FLC follows a periodic pattern of expression with reactivation occurring on a yearly basis.

Although numerous attempts have been made to characterize environmentally induced epigenetic changes in plants, the inheritance of such variations has not been systematically investigated. In general, both the altered epigenetic marks and the corresponding phenotypic modifications are transient and thus reversal occurs after the inducing stimulus is removed. However, it seems in some instances that stress-induced epigenetic polymorphism can be transmitted to nonexposed offspring. The occurrence in these studies of either preexisting genetic polymorphisms or stress-induced mutations has not always been ruled out as bases for the heritable traits. This issue is especially important when considering the work by Molinier *et al.* (2006), who used a potentially mutagenic ultraviolet (UV) treatment as a stress condition, and who provided evidence for the impact of genetic origin in the transgenerational inheritance of an epigenetic trait. It must be noticed, in addition, that in most of these cases, the transmission of epigenetic alterations has only been assessed for the F1 generation after stress only, so there is a possibility that the stress-induced relaxation of the mechanisms aimed at safeguarding genome stability has allowed an unusually high degree of "carry-over" of epigenetic marks from the stressed plant to its nonstressed progeny. Such a parental effect is not considered as actual transgenerational epigenetic inheritance, since it is stochastic and unstable. To date, the most solid demonstration of transgenerational inheritance has been given by Kou *et al.* (2011), who have followed the transmission of nitrogen-deficiency-induced DNA methylation polymorphisms (paralleled by enhanced tolerance to N deprivation) through three successive generations in a selfing pure-line cultivar of rice. It is likely that in the coming years, more instances of true inheritance of environmentally induced traits will be unraveled and that this will help to discriminate between random, transient variation resulting from stress and targeted adaptive regulation mechanisms yielding heritable phenotypes. Ultimately, shedding light on the possible contribution of acquired epigenetic variation to transgenerational stress tolerance and acclimation could potentially enable plant breeders to design progressive selection schemes aimed at optimizing growth and production for a given set of agroecological conditions (Rival and Jaligot 2011).

Epigenetic Control of Plant Development

Successful sequencing of model plants with well-annotated genomes, such as *Arabidopsis* (Kawakatsu *et al.* 2016), rice (Tang *et al.* 2018), and soybean (Schmitz *et al.* 2013), has created more opportunities for exploring the epigenome, including whole-genome methylation analysis at single-nucleotide resolution. Comprehensive maps of DNA methylation patterns have allowed us to broaden the knowledge and understanding of the potential tissue-specific epigenetic variations and functions in plants. Widman *et al.* (2014) compared DNA methylation, nucleosome distributions, and transcriptional levels in the shoots and roots of *Arabidopsis* accession Columbia-0 (Col-0) and linked the observed organ-specific alterations in gene activity to particular epigenetic profiles. Despite the lack of global variations in DNA methylation levels between the studied tissues, hypermethylated genome regions tend to occur preferentially in shoots relative to roots. The same authors identified a group of genes belonging to the extensin family that have at least tenfold higher expression and lower nucleosome density in roots relative to shoots. Earlier studies of DNA methylation and demethylation patterns in four rice genotypes indicated that the relative DNA hypomethylation in roots ensure greater plasticity and adaptability to stress (Karan *et al.* 2012). Although other studies have also demonstrated tissue-specific relationships between various chromatin modulators and gene expression, this topic is still a subject of debate. Specialized tissues, such as embryo, endosperm, and pollen, have shown large-scale changes in the expression of specific genes and DNA methylation patterns during *Arabidopsis* development. In pollen and somatic cells, the methylation is maintained by comparable mechanisms; however, higher efficiency of CG methylation maintenance has been noted in pollen, which could contribute to the inheritance of methylation across generations (Hsieh *et al.* 2016).

Epigenetic Control of Plant Response to Abiotic Stress

Plants are constantly exposed to environmental stresses, including low water and nutrient availability, extreme temperatures and light intensities, and soil properties such as salinity and heavy metal content. To cope with climate change and resulting increasing occurrences of unpredictable environmental conditions, plants have developed genetic and epigenetic mechanisms that enable them to withstand single or combined stresses and their interactions (Shanker and Venkateswarlu 2011). Thus, understanding this complexity in crops requires knowledge of the genetic and epigenetic bases of responses to environmental changes. In one example of such research, Brzezinka *et al.* (2016) used priming by heat stress as a model to dissect the memory of environmental stresses in Arabidopsis in order to identify genes that are specifically required for heat stress memory, but not for the initial responses to heat. The authors identified the FORGETTER1 (FGT1) gene, and found that the FGT1 protein binds directly to a specific class of heat-inducible genes and ensures that the heat-inducible genes are always accessible and active by modifying the way the DNA containing these genes is packaged. Their findings could lead to new approaches in crop breeding programs for enhancing the resistance to abiotic stress, as knowledge of stability and heritability features of epigenetic marks and epigenetic regulatory mechanisms is crucial for breeding applications (Gallusci *et al.* 2017).

Exploiting Epigenetic Diversity

Natural Epialleles

Naturally occurring epialleles have been associated with morphological, developmental, metabolic, and immune properties in plants, all of which are significant from an agricultural standpoint. Many of these epialleles appear to have developed as spontaneous epimutations, that is, through stochastic gain or loss of DNA methylation, even if the origin of these epialleles is not always totally evident. Examples in plants include the spontaneous hypomethylation of the rice Fertilization-independent endosperm1 (FIE1) gene, which is linked to shorter stature, and the spontaneous hypermethylation of the tomato colorless non-ripening (CNR) locus, which inhibits fruit maturation. Epialleles offer prospective breeding or supplemental editing targets for the agricultural industry because they are frequently meiotically stable and independent of genetic variation.

Chemically Induced Epigenetic Diversity

Various chemicals that can change epigenetic information are described, and according to their function, they are classified into two main groups: DNMT inhibitors and acetylase inhibitors. The first group consists mainly of a cytidine analog that specifically inhibits DNA methylation by sequestering DNMTs. Once incorporated into DNA during replication, analogous methyltransferases trap DNA and mediate their degradation, resulting in an unpassable loss of DNA methylation in the cell. The most commonly used components are 5-AzaC (5-azacytidine) and zebularine, but there are also more stable variants. The second major group consists of inhibitors of HDAC, which are classified into different groups such as hydroxamic acid, aminobenzamide, cyclic peptide, and short-chain fatty acids (Flowchart 3).

When studying epigenetic mechanisms, investigators are confronted by two main obstacles: (1) the extreme variability of the epigenetic variations between different individual cells, tissues, developmental stages, or growth conditions and (2) the technical restrictions intrinsic to the methods used to characterize these variations, which are based mostly on the simultaneous examination of a limited number of loci (Grunau *et al.* 2001; Haring, *et al.* 2007). The emergence and the ever-increasing affordability of high-throughput sequencing technologies have promoted a spectacular scaling-up in the epigenetic studies undertaken in recent years (Morozova *et al.* 2008). In the wake of the sequencing of model plant genomes (*A. thaliana* and rice), genome-wide methylation mapping has been performed by combining bisulfite conversion of cytosines with either microarray hybridization (Vaughn *et al.* 2007), or direct sequencing [MethylC-Seq (MethylC-sequencing)]. MethylC-Seq allows the detection of genome-wide DNA methylation patterns at single-nucleotide levels in all sequence contexts. This is particularly important in plants, since in these organisms methylation of cytosines in any sequence context can be achieved through the combined action of three distinct DNMTs families. Additionally, the use of NGS technologies (Varshney *et al.* 2009) such as 454-pyrosequencing or the more recent Solexa/Illumina technology has provided several 100-fold coverages for any locus. This so-called "deep sequencing" allows evaluation of the proportion of DNA fragments displaying a given methylation pattern within a tissue and therefore provides valuable information regarding the influence of cell-specific methylation differences on differential gene expression (Taylor *et al.* 2007).

Flowchart 3: Epigenetic variations that could be used in plant breeding. (5-AzaC: 5-azacytidine; CRISPR: clustered regularly interspaced short palindromic repeats; TALENs: transcription activator-like effector nucleases; ZNFs: zinc finger nucleases)

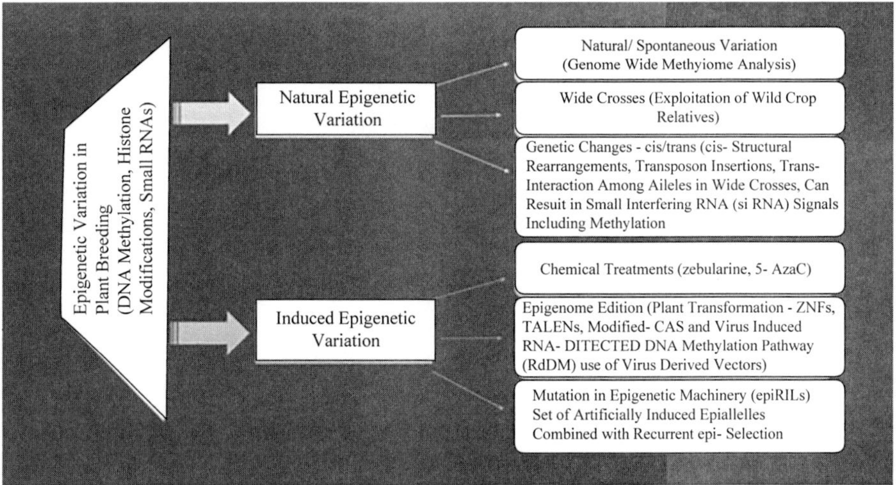

The Rise of Epigenomics: Impact of Next-Generation Sequencing on Epigenetics

In parallel, chromatin immunoprecipitation (ChIP), which is the method of choice to identify the histone modifications surrounding a given DNA sequence, has been taken to a higher level through its coupling with hybridization (ChIP–chip) on tiling microarrays representing whole chromosomes or genomes. More recently, NGS technologies have enabled ChIP sequencing (ChIP-seq), which combines conventional ChIP followed by direct sequencing (Schones and Zhao 2008). This latter method has been established for the analysis of the human epigenome and several reports are now available for plant systems. The analysis of whole transcriptomes and whole small RNA contents within specific tissues have also greatly benefited from these technical advances. Through the aggregation of the different layers of epigenomic information constituted by the methylome, genome-wide histone modifications, the transcriptome (mRNA-seq), and the small RNAome (smRNAseq), highly integrated epigenome maps have been obtained in *Arabidopsis*, in rice and in maize. These maps, which can be considered as a modern take on the original concept of "epigenetic landscape" coined by Conrad Waddington, give an impressive overview of the coordination between epigenetic and gene expression profiles throughout the genome (Schmitz and Zhan 2011)

Advances of Epigenetics in Plant Breeding

The relevance of epigenetic regulation to crop breeding has been demonstrated for some crops, for example, by its effect on growth vigor and yield in tomato (Yang *et al.* 2015). Silencing of the MutS HOMOLOG1 (MSH1) gene in tomato using RNAi results in enhanced plant growth and productivity, even in the absence of the transgene. Total fruit weight and number are increased under field conditions. In addition, under high-temperature field conditions, the MSH1-silenced line produced a higher proportion of red ripe fruits, similarly to the FLA8044 heat-tolerant cultivar. These phenotypic changes are linked to DNA methylation, as the methylation inhibitor 5-AzaC represses the observed phenotypes. In a later study, it was shown that Methyltransferase 1 (MET1) and histone deacetylase 6 (HDA6) are essential components of these changes (Kundariya *et al.* 2020). In *Arabidopsis*, MSH1 mutants displayed enhanced tolerance to drought and salt stress, and increased susceptibility to freezing temperatures. This example, along with various others (Table 1), highlights the link between epigenetics and multiple important traits as well as the potential of epigenetics in crop breeding. Additional examples demonstrating the importance of epigenetic regulation of crop resilience and productivity to environmental and endogenous factors are anticipated to be uncovered in the coming years.

Table 1: Examples of epigenetics modifications for crop improvement.

Sr. No.	Species	Modification	Main conclusions toward crop improvement	Reference
1	*Arabidopsis*, tomato	Small RNA	They showed how the enhanced plant vigor phenotypes of the MSH1 system are reproducible in tomato field size experiments and therefore demonstrated how epigenetic perturbation strategies can be used in crops.	Kundariya *et al.* 2020
2	Arabidopsis, White clover	DNA methylation	Studies focused on description of DNA methylation in stress memory phenomenon.	Yang *et al.* 2020
3	Canola	DNA methylation and histone modifications	The shaping of the epigenome has the potential to artificially increase yield in crops.	Hauben *et al.* 2009
4	Cotton	DNA methylation	DNA methylation is suggested to affect photoperiodic flowering time and seed dormancy.	Song *et al.* 2017
5	Grapevine	DNA methylation	Conservation of DNA methylation changes in response to medium-to-high temperatures in regenerated plants	Baránek *et al.* 2015
6	Maize	DNA methylation	The methylation map will provide an invaluable resource for epigenetic studies in maize and how methylation patterns can be used to predict key phenotypes.	Regulski *et al.* 2013
7	Maize	DNA methylation	DNA methylation controls cell division in maize leaves and correlates with the mitotic exit and entering cell expansion.	Candaele *et al.* 2014
8	Oil palm	DNA methylation and small RNA	The ability to predict and cull mantling at the plantlet stage will facilitate the introduction of higher-performing clones and optimize environmentally sensitive land resources.	Ong-Abdullah *et al.* 2015
9	Rapeseed	DNA methylation	DNA hypomethylation is required for plant cell reprogramming to initiate microspore embryogenesis and doubled haploid production for crop breeding.	Solís *et al.* 2012
10	Tomato	DNA methylation	Multiple changes in gene methylation were linked to the ethylene pathway and ripening processes.	Zuo *et al.* 2020

11	Wheat	DNA methylation and chromatin modifications	Epigenetic modifications contribute to the expression divergence of three TaEXPA1 homoeologs during wheat development.	Hu *et al.* 2013
12	Tobacco	DNA methylation	Aluminum stress, salt, and low temperature treatments induced demethylation patterns. These results suggested a close correlation between methylation and expression of NtGPDL upon abiotic stresses with a cause–effect relationship.	Choi and Sano 2007.
13	Soybean	DNA methylation and histone modifications	Salinity stress was shown to affect the methylation status of several transcription factors (one MYB, one b-ZIP, and twoAP2/DREB family members). For some of them, DNA methylated transcription factors were correlated with an increased level of histone H3K4 trimethylation and H3K9 acetylation, and/or a reduced level of H3K9 demethylation in various parts of the promoter or coding regions.	Song *et al.* 2012
14	Sugar beet	DNA methylation	Tolerance to bolting is an agronomic trait for biennial cultivated sugar beet. Bolting is associated with the use of sucrose root stock and should be avoided in the field. Here, tolerance to bolting was correlated to epigenomic polymorphism in DNA methylation, notably in genes involved in cold acclimation, hormonal pathway genes, and flowering genes.	Hébrard *et al.* 2016
15	Rice	DNA methylation	Multi-generational drought improves drought adaptability of offspring, which could be linked to non-random appearance of drought-induced transgenerational epimutations. Some of the genes related to these epimutations are directly involved in stress-responsive pathways.	Zheng *et al.* 2017

Conclusion

Plant breeding relies on the variation of important traits and the effectiveness of selecting these traits. Epigenetic differences can lead to inheritable traits, and epimutations generate new variations faster than DNA mutations, making them a promising source of trait variations, especially in breeding populations with limited genetic diversity. Epigenetics is a powerful tool in plant breeding, offering the potential to enhance crop performance and sustainability. By understanding and manipulating epigenetic modifications, breeders can improve traits that are challenging to achieve through traditional methods. This includes increasing yield potential, improving nutrient uptake efficiency, enhancing stress tolerance, and reducing susceptibility to pests and diseases. Epigenetic breeding also provides an alternative to genetically modified organisms (GMOs) in agriculture. However, further research is necessary to fully comprehend the interactions between genes, the environment, and epigenetic modifications. Epigenetic mechanisms have significant impacts on phenotypic traits and can be applied in plant breeding, forestry, and agriculture. However, there are still many unexplained plant epigenetic mechanisms, limiting their biotechnological applications. Future research should focus on understanding epigenetic regulation in non-model species to enable new applications in crop improvement and adaptation to climate change.

In the future, breeders must consider crop epialleles and their role in adapting to environmental changes. Understanding transgenerational epigenetic inheritance can lead to the development of epialleles tailored for specific environmental conditions. However, more research is needed on a wider range of plant species to fully comprehend the mechanisms behind epigenetic variation in crops. This requires collaboration among researchers in different areas of plant science and better integration of epigenomic data from different crops.

References

Baránek, M., Čechová J., Raddová J., Holleinová V., Ondrušíková E., Pidra M. "Dynamics and Reversibility of the DNA Methylation Landscape of Grapevine Plants (Vitis vinifera) Stressed by In Vitro Cultivation and Thermotherapy." PLoS One 10, 5 (2015):e0126638.

Boyko, A., Kovalchuk I. "Genome Instability and Epigenetic Modification—Heritable Responses to Environmental Stress?" Current Opinion in Plant Biology 14, 3 (2011):260-6.

Bräutigam, K., Vining K.J., Lafon Placette C., Fossdal C.G., Mirouze M., Marcos J.G., et al. "Epigenetic Regulation of Adaptive Responses of Forest Tree Species to the Environment." Ecology and Evolution 3, 2 (2013):399-415.

Brzezinka, K., Altmann S., Czesnick H., Nicolas P., Gorka M., Benke E., et al. "Arabidopsis FORGETTER1 Mediates Stress-induced Chromatin Memory through Nucleosome Remodeling. elife 5 (2016):e17061.

Candaele, J., Demuynck K., Mosoti D., Beemster G.T., Inzé D., Nelissen H. "Differential Methylation During Maize Leaf Growth Targets Developmentally Regulated Genes." Plant Physiology 164, 3 (2014):1350-64.

Choi, C.S., Sano H. "Abiotic-stress Induces Demethylation and Transcriptional Activation of a Gene Encoding a Glycerophosphodiesterase-like Protein in Tobacco Plants." Molecular Genetics and Genomics 277 (2007):589-600.

Feil, R. "Epigenetics, an Emerging Discipline with Broad Implications." Comptes Rendus Biologies 331, 11 (2008):837-43.

Finnegan, E.J. "Epialleles—A Source of Random Variation in Times of Stress." Current Opinion in Plant Biology 5, 2 (2002):101-6.

Gallusci, P., Dai Z., Génard M., Gauffretau A., Leblanc-Fournier N., Richard-Molard C., et al. Epigenetics for Plant Improvement: Current Knowledge and Modeling Avenues." Trends in Plant Science 22, 7 (2017):610-23.

Grunau, C., Clark S.J., Rosenthal A. "Bisulfite Genomic Sequencing: Systematic Investigation of Critical Experimental Parameters." Nucleic Acids Research 29, 13 (2001):e65.

Hall, B.K. "In search of Evolutionary Developmental Mechanisms: The 30-year Gap between 1944 and 1974." Journal of Experimental Zoology Part B: Molecular and Developmental Evolution 302, 1 (2004):5-18.

Haring, M., Offermann S., Danker T., Horst I., Peterhansel C., Stam M. "Chromatin Immunoprecipitation: Optimization, Quantitative Analysis and Data Normalization." Plant Methods 3, 1 (2007):11.

Hauben, M., Haesendonckx B., Standaert E., Van Der Kelen K., Azmi A., Akpo H., et al. "Energy Use Efficiency Is Characterized By An Epigenetic Component that Can Be Directed through Artificial Selection to Increase Yield." Proceedings of the National Academy of Sciences 106, 47 (2009):20109-14.

He, G., Elling, A.A., Deng X.W. "The Epigenome and Plant Development." Annu. Rev. Plant Biol. 62 (2011):411-35.

Hébrard, C., Peterson D.G., Willems G., Delaunay A., Jesson B., Lefèbvre M., et al. "Epigenomics and Bolting Tolerance in Sugar Beet Genotypes." Journal of Experimental Botany 67, 1 (2016):207-25.

Hsieh, P.H., He S., Buttress T., Gao H., Couchman M., Fischer R.L., et al. "Arabidopsis Male Sexual Lineage Exhibits More Robust Maintenance of CG Methylation than Somatic Tissues." Proceedings of the National Academy of Sciences 113, 52 (2016):15132-7.

Hu, Z., Han Z., Song N., Chai L., Yao Y., Peng H., et al. "Epigenetic Modification Contributes to the Expression Divergence of Three T a EXPA 1 Homoeologs in Hexaploidy Wheat (Triticum aestivum)." New Phytologist 197, 4 (2013):1344-52.

Kakoulidou, I., Avramidou E.V., Baránek M., Brunel-Muguet S., Farrona S., Johannes F., et al. "Epigenetics for Crop Improvement in Times of Global Change." Biology 10, 8 (2021):766.

Karan, R., DeLeon T., Biradar H., Subudhi P.K. "Salt Stress Induced Variation in DNA Methylation Pattern and Its Influence on Gene Expression in Contrasting Rice Genotypes." PloS One 7, 6 (2012):e40203.

Kawakatsu, T., Huang S.S.C., Jupe F., Sasaki E., Schmitz R.J., Urich M.A., et al. "Epigenomic Diversity in A Global Collection of Arabidopsis thaliana Accessions." Cell 166, 2 (2016):492-505.

Klemm, S.L., Shipony Z., Greenleaf W.J. "Chromatin Accessibility and the Regulatory Epigenome." Nature Reviews Genetics 20, 4 (2019):207-20.

Kou H.P., Li Y., Song X.X., Ou X.F., Xing S.C., Ma J. "Heritable Alteration in DNA Methylation Induced by Nitrogen-deficiency Stress Accompanies Enhanced Tolerance by Progenies to the Stress in Rice (Oryza sativa L.) Journal of Plant Physiology 168, 14 (2011):1685-93.

Kundariya, H., Yang X., Morton K., Sanchez R., Axtell M.J., Hutton S.F., et al. "MSH1-induced Heritable Enhanced Growth Vigor through Grafting is Associated with the RdDM Pathway in Plants." Nature Communications 11, 1 (2020):5343.

Meccariello, R. 2019. Introductory Chapter: Epigenetics in Summary. Epigenetics. IntechOpen. doi:10.5772/intechopen.86541.

Molinier, J., Ries G., Zipfel C., Hohn B. "Transgeneration Memory of Stress in Plants." Nature 442, 7106 (2006):1046-9.

Moore, L.D., Le T., Fan G. "DNA Methylation and its Basic Function." Neuropsychopharmacology 38, 1 (2012):23-38.

Morange. M. "La Tentative de Nikolai Koltzoff (Koltsov) de Lier Génétique, Embryologie et Chimie Physique." Journal of Biosciences 36, (2011):211-4.

Morozova, O., Marra M.A. "Applications of Next-generation Sequencing Technologies in Functional Genomics." Genomics 92, 5 (2008):255-64.

Ong-Abdullah, M., Ordway J.M., Jiang N., Ooi S.E., Kok S.Y., Sarpan N., et al. "Loss of Karma Transposon Methylation Underlies the Mantled Somaclonal Variant of Oil Palm." Nature 525, 7570 (2015):533-7.

Quadrana, L., Colot V. "Plant Transgenerational Epigenetics." Annual Review of Genetics 50 (2016):467-91.

Rabajante, J.F., Babierra A.L. "Branching and Oscillations in the Epigenetic Landscape of Cell-fate Determination." Progress in Biophysics and Molecular Biology 117, 2-3 (2015):240-9.

Regulski, M., Lu Z., Kendall J., Donoghue M.T., Reinders J., Llaca V., et al. "The Maize Methylome Influences mRNA Splice Sites and Reveals Widespread Paramutation-like Switches Guided by Small RNA." Genome Research 23, 10 (2013):1651-62.

Samo, N., Ebert A., Kopka J., Mozgová I. "Plant Chromatin, Metabolism and Development–an Intricate Crosstalk." Current Opinion in Plant Biology 61 (2021):102002.

Schmitz, R.J., Zhan X. "High-throughput Approaches for Plant Epigenomic Studies." Current Opinion in Plant Biology 14, 2 (2011):130-6.

Schmitz, R.J., He Y., Valdés-López O., Khan S.M., Joshi T., Urich M.A., et al. "Epigenome-wide Inheritance of Cytosine Methylation Variants in a Recombinant Inbred Population." Genome Research 23, 10 (2013):1663-74.

Schones, D.E., Zhao K. "Genome-wide Approaches to Studying Chromatin Modifications. Nature Reviews Genetics 9, 3 (2008):179-91.

Shanker, A. Venkateswarlu B. 2011. Abiotic Stress in Plants—Mechanisms and Adaptations. InTech. doi:10.5772/895.

Singh, J., Mishra V., Wang F., Huang H.Y., Pikaard C.S. "Reaction Mechanisms of Pol IV, RDR2, and DCL3 Drive RNA Channeling in the siRNA-Directed DNA Methylation Pathway." Molecular Cell 75 (2019):576-89.e5.

Solís, M.T., Rodríguez-Serrano M., Meijón M., Cañal M.J., Cifuentes A., Risueño M.C., et al. "DNA Methylation Dynamics and MET1a-like Gene Expression Changes During Stress-induced Pollen Reprogramming to Embryogenesis." Journal of Experimental Botany 63, 18 (2012):6431-44.

Song, Q., Zhang T., Stelly D.M., Chen Z.J. "Epigenomic and Functional Analyses Reveal Roles of Epialleles in the Loss of Photoperiod Sensitivity During Domestication of Allotetraploid Cottons." Genome Biology 18 (2017):1-14.

Song, Y., Ji D., Li S., Wang P., Li Q., Xiang F. "The Dynamic Changes of DNA Methylation and Histone Modifications of Salt Responsive Transcription Factor Genes in Soybean." PloS One 7, 7 (2012):e41274.

Sung, S., Amasino RM. "Vernalization and Epigenetics: How Plants Remember Winter. Current Opinion in Plant Biology 7, 1 (2004):4-10.

Talbert, P.B., Henikoff S. "Histone Variants at a Glance." Journal of Cell Science 134 (2021):jcs244749.

Tang, X., Liu G., Zhou J., Ren Q., You Q., Tian L., et al. "A Large-scale Whole-genome Sequencing Analysis Reveals Highly Specific Genome Editing by both Cas9 and Cpf1 (Cas12a) Nucleases in Rice." Genome Biology 19, 1 (2018):1-13.

Taylor, K.H., Kramer R.S., Davis J.W., Guo J., Duff D.J., Xu D. "Ultradeep Bisulfite Sequencing Analysis of DNA Methylation Patterns in Multiple Gene Promoters by 454 Sequencing." Cancer Research 67, 18 (2007):8511-8.

Tirnaz, S., Batley J. "Epigenetics: Potentials and Challenges in Crop Breeding." Molecular Plant 12, 10 (2019):1309-11.

Varotto, S., Tani E., Abraham E., Krugman T., Kapazoglou A., Melzer R., et al. "Epigenetics: Possible Applications in Climate-Smart Crop Breeding." Journal of Experimental Botany 71, 17 (2020):5223-36.

Varshney R.K., Nayak S.N., May G.D., Jackson S.A. "Next-generation Sequencing Technologies and their Implications for Crop Genetics and Breeding." Trends in Biotechnology 27, 9 (2009):522-30.

Vaughn M.W., Tanurdzic M., Lippman Z., Jiang H., Carrasquillo R., Rabinowicz P.D. "Epigenetic Natural Variation in Arabidopsis thaliana." PLoS Biology 5, 7 (2007):e174.

Waddington C.H. "The Epigenotype." Endeavour 1 (1942):18-20.

Waddington C.H. 2014. The Epigenetics of Birds. Cambridge: Cambridge University Press.

Waddington, C.H. 1940. Organisers and Genes. Cambridge: The University Press.

Wang Y.M., Lin X.Y., Dong B., Wang Y.D., Liu B. "DNA Methylation Polymorphism in a Set of Elite Rice Cultivars and its Possible Contribution to Inter-cultivar Differential Gene Expression." Cellular & Molecular Biology Letters 9, 3 (2004) 543-56.

Widman, N., Feng S., Jacobsen S.E., Pellegrini M. "Epigenetic Differences between Shoots and Roots in Arabidopsis Reveals Tissue-specific Regulation." Epigenetics 9, 2 (2014):236-42.

Yang, X., Sanchez R., Kundariya H., Maher T., Dopp I., Schwegel R., et al. "Segregation of an MSH1 RNAi Transgene Produces Heritable Non-genetic Memory in Association with Methylome Reprogramming." Nature Communications 11, 1(2020):2214.

Yang, X., Kundariya H., Xu Y.Z., Sandhu A., Yu J., Hutton S.F., et al. "MutS HOMOLOG1-Derived Epigenetic Breeding Potential in Tomato." Plant Physiology 168 (2015):222-32.

Zheng, X., Chen L., Xia H., Wei H., Lou Q., Li M., et al. "Transgenerational Epimutations Induced by Multi-generation Drought Imposition Mediate Rice Plant's Adaptation to Drought Condition." Scientific Reports 7, 1 (2017):39843.

Zuo, J., Grierson D., Courtney L.T., Wang Y., Gao L., Zhao X., et al. "Relationships between Genome Methylation, Levels of Non□coding RNAs, mRNAs and Metabolites in Ripening Tomato Fruit." The Plant Journal 103, 3 (2020):980-94.

16

Concepts and Models of Stability Analysis in Plant Breeding

S. Sasipriya[1], K. Divya[2] and Dharm Veer Singh[3]

[1]Department of Genetics and Plant Breeding, Malla Reddy University Hyderabad, India

[2]Department of Genetics and Plant Breeding, PJTSAU, Hyderabad, India

[3]Department of Genetics and Plant Breeding, Banda University of Agriculture and Technology, Banda, Uttar Pradesh, India

Abstract

An average performance of a variety depends on its genotype worth, environmental conditions, and the effect of interaction between genotype and environment. Stability of a genotype is determined by its mean performance over different locations and generations which has to be confirmed with data from various environmental conditions. A single experimental location can give idea about only mean performance of a variety responding to a specific environment. There are many different stability analyses developed to estimate the average adaptation of a variety or genotype. Among them regression-based models and multiplicative models are found to have more implications in plant breeding. The present book chapter overlooks into various stability models adopted in crop improvement programs.

Introduction

Breeders expect a new variety to show high performance for yield and its contributing characters along with other physiological and nutritional traits. A variety is said to be superior not only for its higher economic yield, but also for its consistent performance over a wide range of environmental conditions. But the performance of a particular variety is the resultant of its genetic constitution and the environment in which it has been grown. Particular variety may not exhibit same phenotypic performance under different environments, or different varieties may perform differently to a specific environment.

The basic cause of differences in the stability of different genotypes is the interaction effect of genotype and environment, known as genotype

× environment (G × E) interaction (GEI). The term environment refers to the uncontrolled external conditions that affect expression of genes of an individual genotype such as the annual rainfall, dry spells, soil fertility gradients, waterlogging, depth and texture of soil, and photoperiod and incidence of disease or pests. However, prediction of GEI is seldom possible with available and known factors. Therefore, the environmental factors can be categorized as predictable and unpredictable environmental factors (Table 1).

Table 1: Predictable and unpredictable factors.

Predictable factors	Unpredictable factors
These are permanent factors of environment and do not fluctuate over generations.	These factors are inconsistent and fluctuate over the generation.
These factors are under human control.	These are uncontrolled factors.
Examples: Soil type, planting date, row spacing	Examples: Rainfall, temperature, relative humidity

Genotypes respond differently across a range of environments, i.e., the relative performance of varieties depends on the environment. The phenotypic mean, i.e., *P*, of a given genotype from the trial may be expressed as:

$$P = \mu + G + E + GE$$

where,

μ is the general population mean, G is the effect of genotype, E is the effect of environment, and GE denotes the interaction between genotype and environment. Presence of GEI causes difficulty in identifying superior genotypes.

Genotype–Environment Interaction

The differential response of genotypes from one environment to another is known as Genotype by Environment interaction. The GEI signifies the relative performance of various genotypes affected by the environment. It reduces the correlation between phenotypic and genotypic values, and has been shown to reduce progress from selection. This adjustment of variety in response to fluctuations in environment for agronomically important traits is called buffering capacity. The variety with high stability is said to be well-buffered.

Patterns of the genotype behaviors in different environments:

- Null GEI
- Simple GEI
- Complex GEI

Significance of GEI

- Identify differences in environmental variation
- Subdivision of geographic region into relatively homogeneous subregions
- Actual genotypic performance
- Deciding breeding strategies

Detection of GEI

- Multi-location replicated trials must be conducted for several years
- Homogeneity test for experimental error using Bartlett's test
- Pooled analysis
- Find genotype and G × E significance from pooled ANOVA (analysis of variance)
- Stability analysis using statistical models

Stability Analysis

Stability refers to the average performance of a genotype with respect to changing environmental factors overtime within given location. Selection of a genotype for stability is possible only when a biometrical model equipped with suitable parameters is at hand to furnish the necessary criteria to rank for stability. A stable genotype will show constant performance irrespective of the environmental conditions, i.e., it has low contribution to the GEI. For this stability analysis to perform generally designs used are randomized complete blocks and incomplete blocks. In case of incomplete blocks, if very large number of genotypes are being estimated lattice designs are used (Fasahat *et al.* 2015).

Flowchart 1: Different models of stability

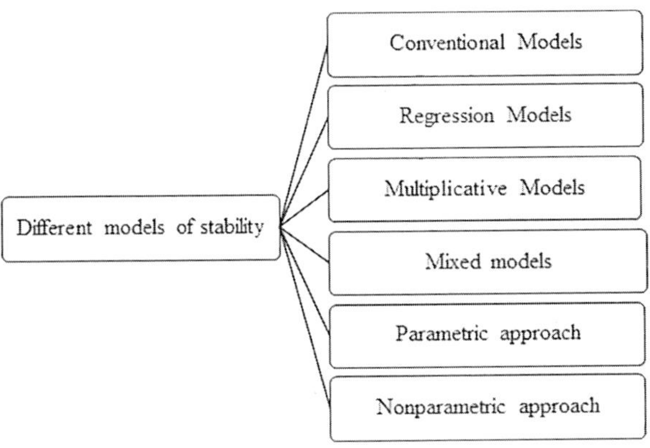

Conventional Models

Analysis of variance is the basic and past model of stability analysis (Elias *et al.* 2016) (Flowchart 2).

$$\gamma_{ijk} = \mu + \alpha_i + \beta_j + (\alpha\beta)_{ij} + \varepsilon_{ijk}$$

where,

γ_{ijk} = yield response variable; μ = overall mean; α_i = genotypic effect; β_j = jth environmental effect; $(\alpha\beta)_{ij}$ = interaction effect; ε_{ijk} = residual error

This model quantifies magnitude of classifiable main effects and interactions; it does not provide the pattern of interactions. It requires replications which may be a constraint.

Flowchart 2: Conventional models

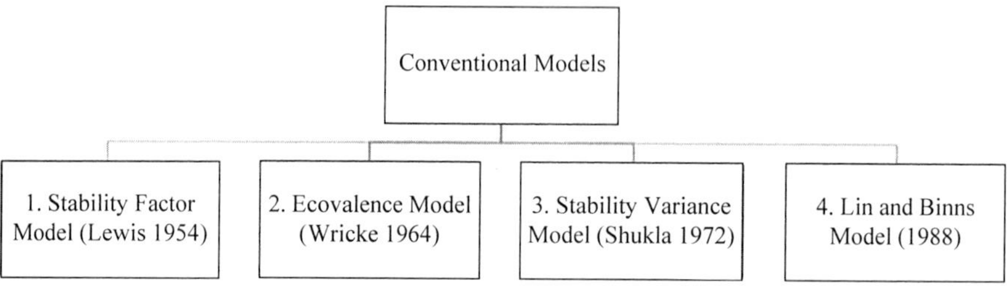

Regression Models of Stability

Average functional relationship between two or more characters is known as regression. It provides the cause and effect situation and also provides the relationship between the dependent and independent variables. It also helps in giving the results of stability across environments (Fig. 1).

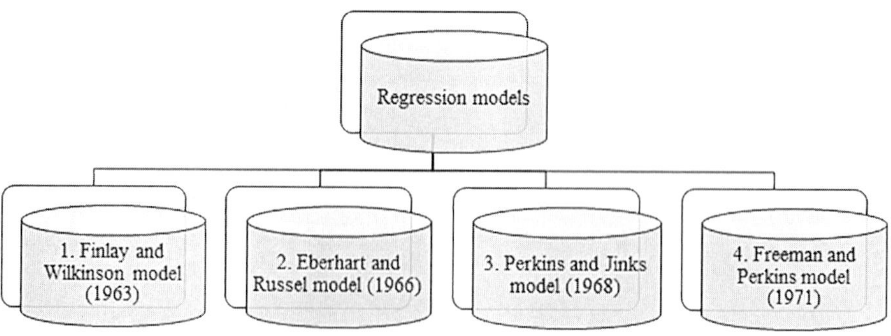

Fig. 1: Regression models

1. ***Finlay and Wilkinson model (1963)***: The model uses two parameters: mean performance over environments and regression coefficient of performance in different environments. Regression coefficient is the linear sensitivity and deviation from average sum of squares due to nonlinear sensitivity which can help in studying the adaptability of different genotypes over locations with different environmental conditions. Linear regression predicts a value of an independent variable based on the independent input variable, and the sensitivity analysis is a series of mathematical analysis which helps to quantify the uncertainty in the dependent variable based on uncertainty in independent variable.

When the regression coefficient of unity, it indicates average stability and a value of zero indicates absolute stability. A value of regression coefficient is >1 means below average stability and the value of regression coefficient is <1 means above average stability (Fig. 2).

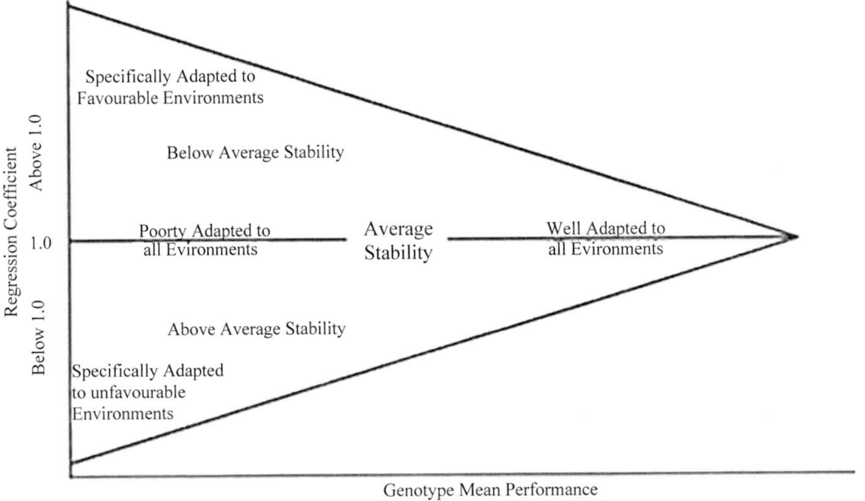

Fig. 2: Genotype regression coefficients plotted against genotype performance.
Source: Adapted from Finlay and Wilkinson 1963.

The analysis of this model is simple as there are only two parameters were used to assess the phenotypic stability such as, mean yield over locations and regression coefficient. However, the deviations from the regression line are not estimated which are important for the stability analysis. Greater emphasis is given on mean performance over environments than regression coefficients.

2. ***Eberhart and Russel model (1966)***: This model includes three parameters such as genotype mean performance (gi), regression value (*bi:* predictable linear response), and deviation from regression (*sdi*: unpredictable nonlinear response). The analysis for stability is quite simple, requiring comparatively lesser area and other resources, making it less expensive. On the other hand, it does not provide independent estimation for mean performance and environmental index. According to this model, a stable variety is one which is having *high mean yield* with a regression coefficient of unity ($b = 1$) and minimum deviation from the regression lines ($s^2d = 0$) (Table 2).

Table 2: Analysis of variance (ANOVA) of Eberhart and Russel model.

Source of variation	**Degrees of freedom**
Genotypes	g-1
E + G*E interaction	g (e-1)
Environment (linear)	1
GE linear	g-1
Pooled deviations	g (e-2)
Genotype-1	e-2
Genotype 2	e-2
Pooled error	ge (r-1)

(E: ??; GE: Gene–environment)

3. ***Perkins and Jinks model (1968)***: Three stability parameters are used *viz.*, mean (Yi), regression coefficient (bi) and deviation from regression (sdi) (Table 3).

Table 3: Analysis of variance (ANOVA) of Perkins and Jinks model

Source of variation	**Degrees of freedom**
Genotypes	g-1
Environment	e-1
Genotype × environment	(g-1)(e-1)
Heterogeneity among regressions	g-1
Remainder	(g-1)(e-2)
Error	ge(r-1)

The degrees of freedom for environment is e-1 which makes the computation complicated for the estimation of stable genotype.

4. ***Freeman and Perkins model (1971)***: Three stability parameters are used, *viz.*, mean (Yi), regression coefficient (bi), and deviation from

regression (sdi). It is a joint regression analysis and environment is completely independent (Table 4).

Table 4: Analysis of variance (ANOVA) of Freeman and Perkins model

Source of variation	Degrees of freedom
Genotypes	g-1
Environment	e-1
Combined regression	1
Residual (1)	e-2
Interaction (G × E)	(g-1)(e-2)
Heterogeneity of regressions	g-1
Residual (2)	(g-1)(e-2)
Error	ge(r-1)

The major advantages and disadvantages of the regression models are (Malosetti *et al.* 2013) as follows:

Advantages

- Variation in response of genotypes to changing environment
- Environment quality in single dimension.
- Maximum retention of G × E interaction signal

Disadvantages

- Fails to explain large portion of G × E
- Assumes linear relationship between G × E and environment means
- The model is not suitable when dealing with extreme environments.
- Less informative where G × E is too complex
- Environment characters is based on single dimension
- ***Multiplicative models***: It explains complex GEIs by assuming the fixed effect. It also eliminates much of noise in data and explains pattern of GEIs. Examples are AMMI (additive main effects and multiplicative interaction) biplot model and GGE (genotype and genotype environment) biplot (Elias *et al.* 2016).
  - *AMMI Model (*Gauch *1985)*: AMMI is a combination of ANOVA and principal components analysis. It enables to:
    - Visualize the GEI
    - Identify genotypes that are adopted to particular environment
    - Identify genotypes that are broadly adapted

- Classify environments into groups
- Measures stability of genotypes

The G × E main effects are extracted using two-way additive ANOVA model which provides the additive mean effects of genotype, environment, and G × E additive effect, while principal component analysis (PCA) is then applied to ANOVA residuals including GEI which provides the multiplicative interaction. The linear model of AMMI (Gollob 1968) is given as follows:

$$\gamma_{ijl} = \mu + G_i + E_j + (\sum \lambda_k \alpha_{ik} \gamma_{jk}) + e_{ij}$$

where,

γ_{ij}: average yield of the ith genotype in j^{th} environment; μ: general mean; G_i and E_j: the effects of the genotype and environment; λ_k: singular value of the kth axis in the PCA; α_{ik}: eigenvector of the ith genotype for the kth axis; γ_{jk}: eigenvector of the jth environment for the kth axis; n: number of principal components used in the model and e_{ij}: average of the corresponding random errors (Table 5).

Table 5: ANOVA of AMMI model

Source	df
TOTAL	(ger- 1)
Treatment	(ge -1)
Genotype	(g -1)
Environment	(e-1)
Interaction IPCA 1 IPCA 2 Residual	(g-1) (e-1)
blocks	(r-1)
error	(r-1) (ge -1)

(**AMMI:** additive main effects and multiplicative interaction; ANOVA: analysis of variance; *df*: ??; IPCA: incremental principle component analysis)

Stability is estimated by AMMI model using AMMI stability value (ASV) and PCA.

$$\text{ASV: } \sqrt{\left[\frac{\text{IPCA 1 Sum of square}}{\text{IPCA 2 Sum of square}} \text{IPCA 1 score}\right]^2 + (\text{IPCA 2 score})^2}$$

The ASV values can be used to interpret stability of a genotype. A lower ASV indicates genotypes with wider adaptability and vice versa. Similarly, if IPCA2 score is near to “0” indicates more stable genotypes while larger values are more responsive to environment and less stable.

Principal component analysis: G × E characterized by interaction PCA as where genotypes and environments can be simultaneously plotted. Principle component 1 (PC1) provides responses of the genotypes that are proportional to environments associated with G × E without change in the range. PC2 provides information about cultivation locations and are not proportional to environments responsible for G × E.

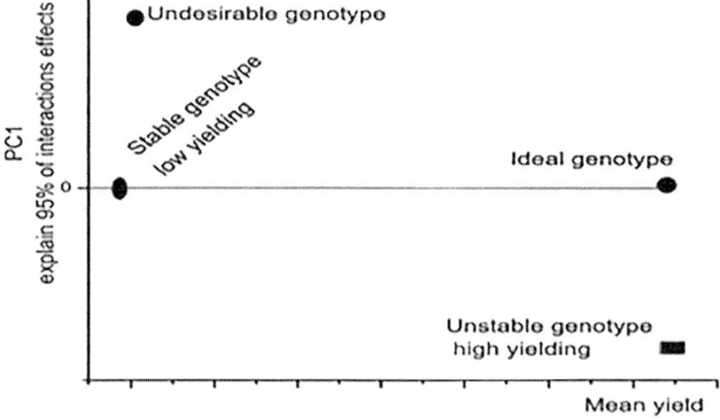

Fig. 3: A biplot graph presenting main types and patterns of stability. (PC1:principle component 1)

Source: Khumairoh *et al.* 2018

The lower the PC value, the higher is the stability of genotype. The AMMI biplot gives the interactions of genotypes and environments either in negative or positive direction. If the angles of vectors from the point of origin are considered and the angle is acute (<90°), it gives positive correlation among environments. If the angle is right angle (90°), there will be no correlation and if the angle is obtuse (>90°) it indicates negative correlation.

- ***Genotype and Genotype Environment biplot:*** In this biplot, ANOVA gives the main effects (G) and PCA gives the interaction effects (G + GE). It addresses three important issues:
  1. Mega-environment evaluation
  2. Genotype evaluation
  3. Test environment evaluation

This method is insensitive to number of genotypes but best predictor for small number of genotypes. It does not explain all the variations present as it does not explain the environment's main effects (Yan *et al.* 2000) (Table 6).

Table 6: Different types of genotype and genotype environment (GGE) biplots.

Raking biplot	**Comparison biplot**	**Joint biplot**
• Shows performance of genotype in specific environment • Gives best genotype • Short lines—high stability • Long lines—low stability	• Compares performance of environment with that of ideal environment • Compares performance of a genotype with that of ideal genotype	• Compares two environments or genotypes simultaneously • Greater the difference from the biplot axis greater is the difference in performance in two environments

Biplots can be created with PCA1 score alone or by including both PCA1 and PCA2. Including both PCA scores for drawing the biplots are more advisable. The genotypes which are present near to the origin of the biplot are considered more stable than one depicted far away from the origin.

Interpretation of biplot

- The PCA values on X and Y axis of the graph divides the graph at the point zero respectively, creating four plots inside the graph.
- The genotypes depicted in the same plot or in a straight line show similar performance and the environments plotted on the same section show similar influence on the genotypes.
- A polygon is formed by joining the genotype which are located far away from the origin encompasses all the genotypes.
- The biplot is divided into different sectors from the origin which help in forming mega-environments. These are having same influence on the genotypes.
- Genotypes plotted near to origin will have more stability and is less likely to be affected by the environmental conditions.

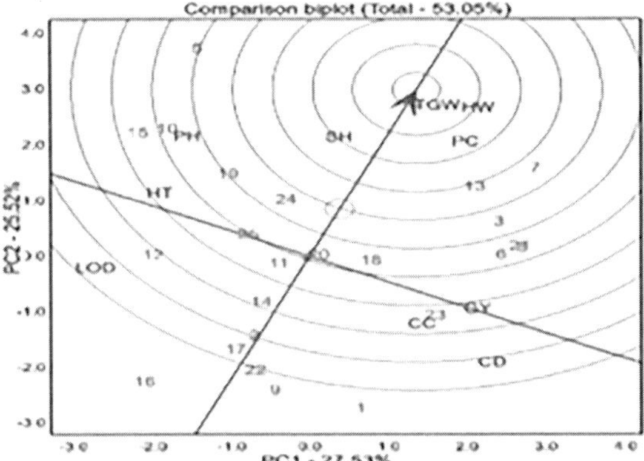

Fig. 4: The relationship genotype by trait in seven environments.

- Comparison biplot helps in comparing different environments and performances of genotypes.
- Smaller the concentric circle, environments share similar attributes with the ideal environments.
- The genotypes in the smaller concentric circle shows high and stable performance.

Source: Oral *et al.*, 2019

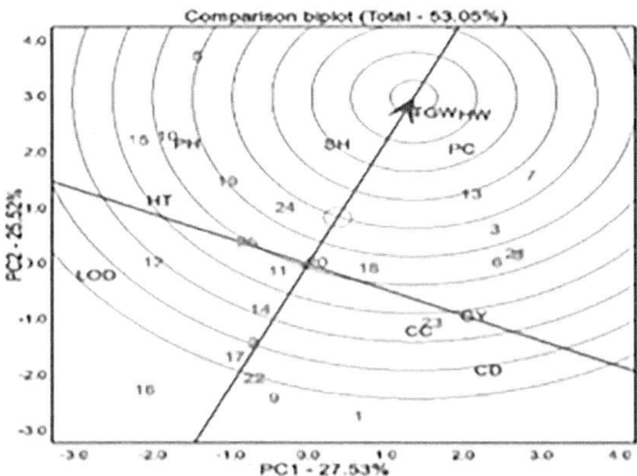

Fig. 5: Ranking genotypes based on traits and instability

- The vector of projection of different genotypes gives an idea about the stability of genotypes.
- If the projected vector is small, genotypes are stable and if the vector is long it is unstable and is sensitive to environment.
- From this biplots, genotype number 7 and 13 can be considered stable.
- Genotypes 5, 10, 15, and 28 are plotted on a long vector, so they are considerably more vulnerable to environmental conditions and are unstable.

Source: Oral *et al.*, 2019

Mixed Models

These models help in estimating the random effects and it is effective for unbalanced data. There is no restriction for replications within environment. It considers heterogeneous variance. BLUP is a mixed model which estimates the stability by using harmonic mean.

***Best linear unbiased estimates (BLUP)*:** Allows to account for environmental factors, even with missing data. The data is shrink towards the mean. It calculates each line for each trait and can make selections based on BLUP. The application of BLUP is main for perennial and horticultural crops. It maximizes the correlation of true genotypic values and predicted genotypic values. There are three methods of assessment in BLUP such as MHGV (harmonic mean of genetic values), RPGV (relative performance of genetic values) and MHPRGV (harmonic mean of relative performance of genetic values).

$$\text{HMGV} = \frac{1}{\sum_{i=1}^{1} \frac{1}{GV_j}} \qquad \text{RPGV} = \frac{1}{1}\left[\frac{\sum GV_j}{M_j}\right]$$

$$\text{HMRPGV} = \sum_{i=1}^{1} \frac{1}{RPGV_j}$$

The linear model of BLUP is, $\mathbf{y = Xm + Zg + Wp + Ti + e}$

Where, y: vector data, m: vector of effects of measurement repeating combinations (assumed to be fixed) added to the overall average, g: vector of the genotypic effects (assumed to be random), p: vector of the permanent environment effect (in this case, random plots), i: vector of the effects of the genotypes × measurements interaction, and e: vector of errors or waste (random).

Advantages of BLUP method

- Provides genotypic values free of environmental factors
- Generates results in its own greatness and character range evaluated
- Effects of G × E eliminates inaccurate information of GEI
- Can be employed in areas including plant and animal breeding

***Recent Models in BLUP*:** Marker effects are estimated either using shrinkage or a combination of shrinkage and variable selection. Commonly used three methods are ridge regression best linear unbiased prediction (RRBLUP), genomic best linear unbiased prediction (GBLUP), and Bayesian models. These models were used in the genomic selection strategy of the crops (Table 7).

Table 7: Recent models in BLUP

RRBLUP	GBLUP	Bayesian models
• Assumption of random and normal distribution with a common variance is made which result in equal shrinkage of the effects. • Markers with large effect on the variable are more tend to be underestimated compared to those having smaller effect. • Appropriate when there are few or no large-effect quantitative trait locus (QTL) and many small-effect QTLs,	• Assumption of multivariate normality • Uses a genomic relationship matrix (GRM) calculated from marker genotypes instead of calculating individual marker effects • Mathematically equivalent to RRBLUP with slight differences in predicting accuracies—attributed to marker—QTL linkage disequilibrium (LD).	• Relax the assumption of common marker effect variances • Allow marker-specific variances • Effectively allowing unequal shrinkage of marker effects • Focused on linkage disequilibrium between marker and QTLs

(**GBLUP:** genomic best linear unbiased prediction; RRBLUP: ridge regression best linear unbiased prediction)

The RRBLUP and GBLUP rely strongly on kinship, if trait variation is dominated by small-effect quantitative trait locus (QTL) than larger effect. In case of Bayesian models, it is considered only if the amount of variation for a trait is large-effect QTL control.

Nonparametric Models

Stability of different genotypes across environments are estimated based on ranks of genotypes. A stable genotype is expected to have same rank over changing environmental conditions. There are no specific assumptions followed in these methods and the interpretation of results are rather easy. These models show reduced sensitivity to measurement errors. Addition or deletion of one or a few observations is not likely to cause great variation in the estimate. The rank of a genotype is obtained by combining mean yield of the genotype and stability. The ranks obtained from these combined estimates are called as adjusted values, which depend only on GEI and error effects. The genotype with the highest adjusted yield is given rank of 1, while the lowest adjusted yield is given a rank of last.

Parametric Models

These models rely on distributional assumptions about genotypic, environmental, and GEI. Good properties under certain statistical assumptions, like normal distribution of errors and interaction effects are taken into consideration. However, the model fails to perform at optimum when assumptions involved gets violated. They are used to assess the stability and adaptability of varieties.

Conclusion

By analyzing the correlations among different stability parameters and giving due consideration to each parameter, breeders can make informed choices when selecting the most suitable method. The key advantage of utilizing stability analysis to choose superior genotypes, as opposed to relying solely on average performance, lies in the reliability of stable genotypes across various environments, which helps reduce the impact of GEI. The GGE biplot offers a comprehensive approach not only for assessing genotypes and environmental factors within a specific season or subsequent seasons, but also for evaluating the consistency of genotypes across multiple years. This method involves consolidating data from various trials conducted in different locations into a single dataset. When there is substantial crossover GEI, it becomes evident that efforts should be directed toward pinpointing location-specific genotypes based on multi-year, multi-location data instead of solely focusing on overall performance. Furthermore, by incorporating information about QTLs present in the genotypes for trait stability analysis, stable genotypes can be identified, and any associated linkage drag can be accounted for. Research indicates that employing stability analyses grounded in various principles can lead to a more accurate identification of stable genotypes.

References

Eberhert, S.A, Russell, W.A. "Stability Parameters for Comparing Varieties." Crop Science 6 (1966):36-40.

Elias, A.A., Robbins K.R., Doerge R.W., Tuinstra M.R. 2016. "Half a Century of Studying Genotype Environment Interactions in Plant Breeding Experiments." Crop Science 56 (2016):2090-105.

Fasahat, P., Rajabi A., Mahmoudi S.B., Noghabi M.A, Rad J.M. "An Overview on the Use of Stability Parameters in Plant Breeding." Biometrics & Biostatistics International Journal 2, 5 (2015):1-11.

Finlay, K.W., Wilkinson, G.N. "The Analysis of Adaptation in a Plant Breeding Program." Australian Journal of Agriculture Research 14, 6 (1963):742-54.

Freeman, G.H., Perkins. "Environmental and Genotype-Environmental Components of Variability VIII. Relation between Genotype Grown in Different Environments and Measurement of These Environments." *Heredity* 27 (1971):15-23.

Gauch Jr, H.G. 1985. Integrating Additive and Multiplicative Models for Analysis of Yield Trials with Assessment of Predictive Success. Ithaca/NY: Cornell University. Mimeo 85-7.

Gollob, H.F. "A Statistical Model which Combines Features of Factor Analysis and Analysis of Variance Techniques." *Psychometrika* 33 (1968):73-115.

Khumairoh, U., Lantinga, E.A. and Schulte, R.P.O. 2018. Complex rice systems to improve rice yield and yield stability in the face of variable weather conditions. *Scientific Reports*. 8: 14746. https://doi.org/10.1038/s41598-018-32915-z.

Lewis, D. "Gene-Environment Interaction. A Relationship between Dominance, Heterosis, Phenotypic Stability and Variability." *Heredity* 8 (1954):333-56.

Lin, C.S., Binns M.R. "A Superiority Performance Measure of Cultivar Performance for Cultivar × Location Data." *Canadian Journal of Plant Sciences* 68 (1988):193-8.

Marcos, M., Jean-Marcel, R. and Fred, A. 2013. The statistical analysis of multi-environment data: modeling genotype-by-environment interaction and its genetic basis. *Frontiers in Physiology*. 4. 10.3389/fphys.2013.00044.

Oral, E., Kendal E., Kilic H., Dogan Y. 2019. "Evolution Barley Genotypes in Multi-Environment Trials by AMMI Model and GGE Biplot Analysis." *Fresenius Environmental Bulletin* 28, 4A (2019):3186-96.

Perkins J.M., Jinks J.L. "Environmental and Genotype-Environmental Components of Variability III. Multiple Lines and Crosses." *Field Crops Research* 46, 1-3 (1968):71-80.

Shukla, G.K. "Some Statistical Aspects of Partitioning Genotype Environmental Components of Variability." *Heredity* 29, 2 (1972):237-45.

Wricke, G. "Zur berechnung der okovalenz bei sommerweizen und hafer." *Z Pflanzenzuchtg* 52 (1964):127-38.

Yan, W., Hunt L.A., Sheng Q, Szlavnics Z. "Cultivar Evaluation and Mega-Environment Investigation Based on the GGE Biplot." *Crop Science* 40, 3 (2000):596-605.

17

Intellectual Property Right and its Requirement in Plant Breeding

***Rinkey Arya*[1], *Sajalsaha*[2], *Deepa Bhadana*[3] *and Rajib Das*[4]**

[1]*Faculty of Agriculture and Agroforestry, DSB Campus Nainital, Kumaun University Uttarakhand, India*

[2]*Ph.D. Nagaland University, School of Agricultural Sciences, Medziphema, Nagaland India*

[3]*Ph.D. Chaudhary Charan Singh University, Meerut, Uttar Pradesh, India*

[4]*Assistant Professor, College of Agriculture, Central Agricultural University Pasighat, Arunachal Pradesh, India*

Abstract

Intellectual property rights (IPRs) are essential for promoting innovation and creativity. They protect intangible creations of the human mind such as inventions, artistic works, and brand names. IPRs grant exclusive rights to creators and inventors, encouraging investment in new ideas and technologies. Patents, copyrights, trademarks, and trade secrets are the main types of IPRs. However, protecting and enforcing IPRs face challenges, including digital piracy and global harmonization. Balancing the rights of intellectual property owners with the public interest is also a challenge. International agreements like Trade-related Aspects of Intellectual Property Rights (TRIPS) establish minimum standards for intellectual property protection, but debates continue on patentability criteria, copyright duration, and public health concerns. It is crucial to strike a balance between exclusive rights and public access to knowledge. Continued dialogue, international cooperation, and adaptive legal frameworks are vital for fostering innovation, economic growth, and societal progress.

Keywords: *Intellectual property, Innovation, Patent, Copyright, Trade secrets.*

Introduction

Intellectual property encompasses the products of human creativity, including inventions, artistic works, and commercial symbols. It can be categorized into two main types: industrial property and copyright. Industrial property covers patents, trademarks, industrial designs, and geographic indications of source. Copyright, on the other hand, protects literary and artistic works such as novels, poems, films, music, and visual arts. Intellectual property rights (IPRs) also extend to related areas such as performers' rights, recording producers' rights, and broadcasting rights. These rights safeguard the interests of creators by granting them exclusive ownership over their creations. However, unlike tangible property, intellectual property lacks physical characteristics and must be expressed in a discernible manner to be protected.

Traditionally, intellectual property comprised four primary types: patents, trademarks, copyrights, and trade secrets. However, the concept has evolved to encompass newer forms, such as geographical indications, plant variety protection, semiconductor and integrated circuit (IC) protection, and undisclosed information. This expansion reflects the changing landscape of innovation driven by rapid technological, scientific, and medical advancements.

Furthermore, in response to global economic shifts, business models now place significant emphasis on the value and growth potential of intellectual property. To comply with international obligations under the World Trade Organization's (WTO) Agreement on Trade-related Aspects of Intellectual Property Rights (TRIPS), India has enacted various legislations to strengthen the protection of IPRs. Intellectual property plays a crucial role in today's world, fostering innovation and facilitating economic development. Its protection is essential for fostering creativity, encouraging investment, and ensuring the advancement of society.

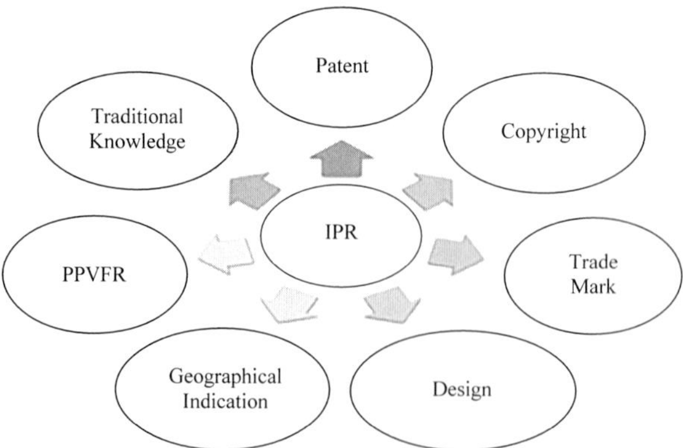

Fig. 1: Classification of Intellectual Property Right (IPR). (PPVFR: Protection of Plant Varieties and Farmers' Rights)

Historical Concept of Intellectual Property Rights

The origin of the intellectual property system can be traced back to renaissance northern Italy, which is considered its birthplace. In 1474, a Venetian law made the initial systematic effort to protect inventions through a form of patent, granting exclusive rights to individuals. Around the same time, Johannes Gutenberg's invention of movable type and the printing press in 1450 contributed to the establishment of the world's first copyright system (Ahuja, 2017).

By the late 19th century, significant advancements in manufacturing techniques fueled large-scale industrialization, accompanied by the rapid growth of cities, expansion of railway networks, increased capital investment, and expanding transoceanic trade. This period witnessed the emergence of stronger centralized governments, industrial ideals, and nationalistic sentiments, leading many countries to enact modern intellectual property laws.

Simultaneously, the international intellectual property system began taking shape with the establishment of the Paris Convention for the Protection of Industrial Property in 1883 and the Berne Convention for the Protection of Literary and Artistic Works in 1886. These conventions laid the foundation for international cooperation in intellectual property protection.

Throughout history, the underlying premise of intellectual property has been that recognizing and rewarding ownership of inventions and creative works stimulates further inventive and creative activity, thereby fostering economic growth. In contemporary corporate settings, ideas and knowledge have become increasingly significant in trade. The value of high-tech products and new medicines largely lies in the level of invention, innovation, research, design, and testing involved.

Films, music recordings, books, computer software, and online services are sought after because of the information and creativity they embody, rather than the materials used to produce them. Even goods that were once considered low-tech commodities now incorporate a higher proportion of invention and design value, such as branded clothing or new plant varieties. Consequently, creators are granted IPRs, allowing them to prevent others from using their inventions, designs, or other creative works.

The concept of intellectual property originated in renaissance Italy and has evolved over time. It has gained prominence as ideas and knowledge have become vital assets in trade, stimulating economic growth. The establishment of international conventions further solidified the intellectual property system, facilitating global cooperation in its protection and enforcement.

The convention establishing the World Intellectual Property Organization (WIPO) (1967) gives the following list of the subject matter protected by IPRs:

- Literary, artistic and scientific works
- Performances of performing artists, phonograms, and broadcasts
- Inventions in all fields of human endeavor
- Scientific discoveries
- Industrial designs
- Trademarks, service marks, and commercial names and designations
- Protection against unfair competition

Intellectual Property Rights System in the World

The intellectual property system has gained further significance worldwide with the establishment of the WTO, which has solidified the importance and role of intellectual property protection through the TRIPS Agreement. Negotiated in 1994 as part of the Uruguay Round of the General Agreement on Tariffs and Trade (GATT) treaty, the TRIPS agreement covers various forms of intellectual property and aims to harmonize and strengthen standards of protection while ensuring effective enforcement at national and international levels.

The TRIPS Agreement encompasses the applicability of general GATT principles and provisions in international agreements on intellectual property (Part I). It sets standards for the availability, scope, and use of IPRs (Part II) as well as their enforcement (Part III), acquisition, and maintenance (Part IV). Additionally, the agreement addresses mechanisms for preventing and settling disputes related to intellectual property (Part V). Formal provisions are outlined in Part VI and VII of the agreement, covering transitional arrangements and institutional frameworks, respectively.

By establishing comprehensive guidelines for intellectual property protection and enforcement, the TRIPS agreement plays a crucial role in promoting innovation, creativity, and economic growth on a global scale.

Intellectual Property System in India

Throughout history, the protection of intellectual property has evolved with the introduction of various legal frameworks. The first known system of intellectual property protection was the Venetian Ordinance in 1485. This was followed by the Statute of Monopolies in England in 1623, which extended patent rights to technology inventions. In the United States, patent laws were established in 1790. European countries developed their own patent laws between 1880 and 1889. In India, the Patent Act was introduced in 1856 and was later modified and amended under "The Indian Patents and Designs Act, 1911", remaining in force for over 50 years. Following India's independence, the comprehensive "Patents Act, 1970" was enacted https://www.indiacode.nic.in.

Traditionally, specific statutes protected only certain types of intellectual output, with copyrights governed by the Copyright Act of 1957, patents by the Patents Act of 1970, trademarks by the Trade and Merchandise Marks (TMM) Act of 1958, and designs by the Designs Act of 1911. However, with India becoming a signatory to the agreement on TRIPS under the WTO, several new legislations were passed to meet international obligations. These included the Trade Marks Act of 1999, replacing the previous legislation on trademarks, the Designs Act of 2000, replacing the 1911 Act on designs, and the Copyright Act of 1957, which has been amended multiple times, with the latest amendment being the Copyright (Amendment) Act of 2012. Additionally, amendments were made to the Patents Act of 1970 in 2005. Geographical indications and plant varieties were also covered by new legislations, namely, the Geographical Indications of Goods (Registration and Protection) Act of 1999 and the Protection of Plant Varieties and Farmers' Rights Act of 2001 (Monagle, 2001).

Over the past 15 years, IPRs have gained prominence in the global economy. In the 1990s, many countries independently strengthened their intellectual property laws, recognizing their significance. At the multilateral level, the agreement on TRIPS within the WTO established a framework for international commitment to the protection and enforcement of IPRs. It is widely acknowledged that stronger IPRs protection promotes innovation and enhances incentives for international technology transfer within the competitive global environment https://ipindia.gov.in/index.htm.

Importance of Intellectual Property Rights

- Encourages innovation and creativity
- Drives economic growth and competitiveness
- Protects intangible assets and competitive advantage
- Facilitates collaboration and technology transfer
- Ensures consumer protection and product quality
- Preserves cultural heritage and diversity
- Facilitates international trade and relations
- Incentivizes investment in research and development
- Drives social and technological progress
- Combats unauthorized use and counterfeiting
- Classification of intellectual property acts in India

Classification of Intellectual Property Rights (Fig. 1)

Patent Act 1970

After India became a signatory to the TRIPS agreement under the WTO, the Patents Act of 1970 underwent several amendments in 1995, 1999, 2002, and 2005 to align with TRIPS obligations. These amendments aimed to harmonize the intellectual property system with international practices while protecting national interests. The amendments also focused on integrating technological advancements in India and ensuring a user-friendly legislation. The Patent Act's rules were subsequently amended in May 2003 and further revised by the Patents (Amendment) Rules of 2005. As a result, the Patent Amendment Act of 2005 is now fully in effect. The head office of the patent office is at Kolkata and its branch offices are located at Chennai, New Delhi, and Mumbai. The Trademarks Registry is at Mumbai and its branches are located in Kolkata, Chennai, Ahmedabad, and New Delhi. The design office is located at Kolkata in the patent office. The offices of the Patent Information System (PIS) and National Institute of Intellectual Property Management (NIIPM) are at Nagpur https://www.wipo.int/patents/en/.

To file a patent in India, you need to follow the procedure outlined below

- ***Determine patentability***: Before filing a patent application, conduct a thorough search to ensure that your invention meets the criteria of novelty, inventive step, and industrial applicability. Assess whether your invention is eligible for patent protection.
- ***Prepare the patent application***: Prepare a detailed description of your invention, including drawings, if necessary. Clearly explain the technical aspects, features, and advantages of your invention. Claims and abstract sections should also be included.
- ***Choose the type of application***: In India, you can file either a provisional or complete patent application. A provisional application allows you to establish an early priority date and provides 12 months' time to file a complete application with all the necessary details.
- ***Filing the application***: Submit the patent application along with the prescribed forms, fees, and supporting documents to the Indian patent office. You can file the application online through the Indian patent office's e-filing system or physically by visiting the patent office.
- ***Publication and examination***: After filing, the patent application is published after 18 months from the filing or priority date. Following publication, a request for examination must be filed within 48 months

from the filing date or priority date. The patent office will examine the application to assess its patentability.

- ***Respond to examination reports***: If any objections or objections are raised during the examination, you need to address them within the stipulated timeframe. Provide suitable responses and amendments to overcome the objections.
- ***Grant of the patent***: If the patent office is satisfied with the examination responses and amendments, they will issue a patent grant. Pay the prescribed fees and complete any formalities required for the grant of the patent.
- ***Maintenance and renewal***: Once the patent is granted, it is essential to maintain it by paying annual renewal fees throughout its validity period. Failure to pay the renewal fees will result in the patent's expiration.

It is important to note that the above steps provide a general overview of the patent filing procedure in India. It is recommended to consult with a qualified patent attorney or agent for specific guidance and to ensure compliance with the requirements and procedures of the Indian patent office.

Trade Marks Act 1999

The Trade Marks Act of 1999 has brought about modernization in trademark law. A trademark is a distinctive symbol used to differentiate the goods of one trader from those of another. In India, trademark protection has been in place for more than four decades under the TMM Act of 1958. India joined the WTO from its inception and became a party to the agreement on TRIPS, which includes IPRs.

In December 1998, India also acceded to the Paris Convention. Recognizing the evolving nature of trading practices, increasing globalization, the importance of investment flows and technology transfer, simplifying trademark management, and incorporating important judicial decisions, the TMM Act of 1958 underwent modernization. To address these goals, the Trademarks Bill was initially introduced in 1994, outlining contemplated changes. However, the bill lapsed in 1994.

Subsequently, a comprehensive review of existing laws was conducted in light of evolving trading practices and increasing globalization. In 1999, the Trademarks Bill was passed by Parliament and received the assent of the President on December 30, 1999, becoming the Trade Marks Act of 1999. This act replaced the TMM Act of 1958, establishing a new framework for trademark protection.

The Trade Marks Act of 1999 reflects the changing landscape of commercial practices and globalization, aiming to align trademark law with contemporary developments and facilitate effective trademark management in India.

Design Act

The Designs Act of 1911 has been replaced by the Designs Act of 2000. The advancement in science and technology necessitated a more efficient legal system to protect industrial designs and encourage design activity. The Designs Act of 2000 aim to provide effective protection to registered designs while promoting the design element in manufactured products. It strikes a balance between the interests involved, ensuring that the law does not grant unnecessary or excessive protection, but rather creates the right incentives for design activity and facilitates the free use of available designs.

The new act is in compliance with the requirements of the agreement on TRIPS, making it relevant for international trade. Industrial design law specifically deals with the aesthetics and originality of an industrial product's design. An industrial product often encompasses both artistic and functional aspects. Design law focuses on protecting the aesthetic appeal of a design while excluding its functional features. For instance, a teacup's functional elements, like the hollow receptacle and handle, cannot be registered. However, any unique and novel ornamental shape or decoration on the cup would be eligible for registration. Similarly, in the case of a table, the flat surface is functional, but aspects such as the shape, color, or leg design fall under the realm of design and can be registered if they are unique and novel.

Overall, the Designs Act of 2000 provides a legal framework to safeguard industrial designs, encourage innovation, and promote aesthetic creativity while respecting functional aspects and the free use of designs.

Geographical Indications of Goods (Registration and Protection) Act

In order to protect the Geographical Indications of goods a Geographical Indications Registry has been established in Chennai to administer the Geographical Indications of Goods (Registration and Protection) Act, 1999 under the Controller general of patents, designs and trade marks (CGPDTM). Previously, there was no provision for registering geographical indications in India, which led to the misuse of Indian geographical indications by individuals outside the country. This misuse involved indicating goods as originating from specific Indian localities when they were not. This issue gained attention, particularly with cases such as the patenting of turmeric, neem, and basmati. It is important to note that under the agreement on TRIPS, reciprocal protection of geographical indications is not obligatory unless the indication is protected in its country of origin.

India lacked specific legislation to effectively protect the interests of producers of goods associated with geographical indications. To address this gap, a comprehensive law called the Geographical Indication of Goods (Registration and Protection) Act, 1999, was passed by the Parliament. This legislation established a framework for the registration and adequate protection of geographical indications. The administration of the law falls under the Geographical Indication Registry, which operates under the supervision of the Controller General of Patents, Designs, and Trade Marks.

The Geographical Indication of Goods Act, 1999, aims to safeguard the authenticity and reputation of Indian geographical indications, preventing unauthorized use and protecting the rights of producers associated with specific regions. Through this legislation, India seeks to provide effective protection and recognition to its unique and valuable geographical indications.

Copyright Act

The Copyright Act of 1957 governs copyright in New Delhi, India. It has been amended over time to stay updated. Copyright protects creative works for the author's lifetime plus 60 years. It covers different types of works like books, music, films, and software. Copyright includes exclusive rights and allows authors to prevent unauthorized use and changes to their work. Piracy is a challenge, especially in the computer industry. Penalties are imposed for using pirated software. Amendments have addressed concerns in industries like books, music, films, computers, and databases.

Copyright protection includes exclusive rights for the author, such as claiming authorship and opposing changes that could harm the creator's reputation. The copyright owner can enforce these rights administratively and in court, including inspecting premises for evidence of piracy and seeking court orders to halt infringing activities and claim damages for financial losses and reputational harm. The computer industry is a significant field that receives copyright protection. The Copyright Act was amended in 1984 to include computer programming as a "literary work". The introduction of a new definition of "computer program" in 1994 broadened the scope of protection to include instructions expressed in any form that can make a computer perform specific tasks or achieve desired results.

Piracy poses significant challenges to the copyright industry, including works such as books, music, films, television programs, computer software, and databases. To address this issue, the Copyright (Amendment) Act of 1994 added a new section, 63B, which imposes penalties for knowingly using an infringed copy of a computer program, including imprisonment and fines. The 1999 Copyright (Amendment) Act introduced provisions allowing purchasers of gadgets/equipment to freely sell them if the item being transacted is not the main item covered under copyright law. This provision enables the fair dealing of broadcasting, especially with the growth of the internet. These amendments demonstrate India's commitment to addressing concerns in the copyright industries, including the book, music, film and television, computer, and database industries.

Protection of Plant Varieties and Farmers' Rights Act, 2001

The Protection of Plant Varieties and Farmers' Rights Act, 2001 is an important legislation in India that aims to protect the rights of both plant breeders and farmers. It recognizes the contributions of farmers in preserving and improving plant genetic resources and provides a framework to safeguard their rights while promoting agricultural innovation and biodiversity conservation (Singh, 2022).

Under this act, farmers are granted specific rights and benefits to protect their interests. Here are some key provisions and benefits for farmers:

- ***Recognition of farmers' rights*:** The act acknowledges and protects the traditional farming practices, knowledge, and contributions of farmers in conserving and improving plant genetic resources.
- ***Benefit-sharing*:** The act emphasizes benefit-sharing arrangements between plant breeders and farmers. Farmers are entitled to receive a share of the royalties or other benefits arising from the commercial exploitation of plant varieties developed from their traditional knowledge (TK) or genetic resources.
- ***Farmers' exemptions*:** The act provides exemptions for farmers to save, use, exchange, share, and sell farm-saved seeds of protected varieties. This enables farmers to continue their customary practices of seed-saving and seed exchange, ensuring access to diverse and locally adapted plant genetic resources.
- ***Recognition of community rights*:** The act recognizes the collective rights of farming communities or tribes over their TK and resources. It enables them to protect and manage their community plant varieties and promote sustainable agricultural practices.
- ***Farmers' rights register*:** The act establishes a Farmers' Rights Register, where farmers can voluntarily register their traditional varieties.

Registration provides legal recognition and protection to these varieties, ensuring their conservation and sustainable use.

- ***Dispute settlement*:** The act establishes mechanisms for the resolution of disputes related to plant varieties and farmers' rights. It ensures that farmers have access to a fair and transparent process for resolving conflicts and protecting their rights.

Overall, the Protection of Plant Varieties and Farmers' Rights Act, 2001, aims to strike a balance between the rights of plant breeders and the rights of farmers. It recognizes and protects the contributions of farmers in agricultural biodiversity conservation and promotes equitable benefit-sharing arrangements. The act empowers farmers, preserves TK, and encourages sustainable agricultural practices, fostering a more inclusive and farmer-centric approach to plant variety protection in India.

Semiconductor Integrated Circuits Layout Design Act, 2000

The electronics and information technology sector is a rapidly growing industry that has made a significant impact on the global economy. Advancements in electronics, computers, and telecommunications have played a crucial role in this growth. Microelectronics, specifically ICs, ranging from small-scale integration to very large-scale integration, are recognized as a core technology in the IT-based society. Protecting the IPRs embedded in IC layout designs is essential to incentivize investments in research and development and drive technological advancements in microelectronics.

Traditional methods of intellectual property protection such as copyright and patents were not adequate for safeguarding layout designs of ICs. The concept of "originality" holds utmost importance in layout designs, regardless of novelty. While patent law requires both originality and novelty, copyright law is too general to accommodate the unique ideas behind scientific creation in IC layout designs. To address this, a sui generis system of protection for layout designs of ICs has been adopted by many countries, usually through separate legislation.

The TRIPS agreement under the WTO includes provisions for establishing standards related to the availability, scope, and use of IPRs, including layout designs of ICs. In line with this, the Semiconductor Integrated Circuit Layout Designs Act was enacted in India in 2000. This act provides for the registration and protection of layout designs of ICs, criteria for distinguishing protectable designs, and rules against registration of nonoriginal or commercially exploited designs, duration of protection, provisions for infringement, payment of royalties, penalties for willful infringement, appointment of a registrar, and an appellate board.

Trade Secrets

Trade secrets consist of confidential business information that gives companies a competitive advantage. They include manufacturing techniques, distribution methods, consumer profiles, advertising strategies, supplier and client lists, and more. Unlike patents, trade secrets do not require registration and can be protected indefinitely. Secrecy is essential, and improper acquisition is prohibited. Protecting TK is crucial for benefiting from it.

Traditional Knowledge

Traditional knowledge refers to ancient and indigenous knowledge transmitted orally across generations. It encompasses diverse areas such as medicinal plant usage, traditional dances, hunting techniques, and crafts. TK is often found in cultural expressions like stories, legends, folklore, rituals, and songs. Historically, there was no protection for TK, but now it is recognized as intellectual property under the TRIPS Agreement. The Indian government has established the Traditional Knowledge Digital Library (TKDL), containing 250,000 formulations from Indian medicine systems (Singh, 2009) https://www.wipo.int/tk/en/.

Intellectual Property Rights Biodiversity

Biodiversity refers to the variety of different life forms within the Earth's Biosphere. It is essential for sustaining life on our planet and plays a vital role in the functioning of ecosystems, providing us with essential products and services. Changes in biodiversity can have significant impacts on human well-being and the well-being of all other living creatures.

The Convention on Biological Diversity (CBD) is an international treaty established in 1992. It was signed at the Earth Summit in Rio de Janeiro and came into effect in December 1993. The CBD aims to protect biodiversity, promote the sustainable use of its components, and ensure the fair sharing of benefits derived from the use of genetic resources. Almost all countries, with 193 parties, have participated in the convention. It addresses various threats to biodiversity and ecosystem services, including those posed by climate change. The convention employs scientific assessments, the development of tools and incentives, technology transfer, and the involvement of stakeholders such as indigenous communities, youth, NGOs, women, and businesses.

The Cartagena Protocol on Biosafety is a supplementary agreement to the CBD. It focuses on safeguarding biological diversity from potential risks associated with living modified organisms resulting from modern biotechnology. The treaty defines biodiversity as the variability among living organisms from all sources, including terrestrial, marine, and other aquatic ecosystems as well as the ecological complexes they are part of. It encompasses diversity within species, between species, and of ecosystems. The Convention reinforces the principle of state sovereignty, granting countries the right to utilize their resources while ensuring that their actions do not harm the environment of other nations. It also establishes a legal framework for regulating access to biological resources and the sharing of benefits arising from their use.

India, as a party to the CBD, has enacted the Biological Diversity Act in 2002. This act aims to conserve biological resources and associated knowledge while facilitating sustainable access to them through a fair and transparent process. In alignment with international commitments, India has also developed legal frameworks concerning intellectual property laws and biodiversity management (Rao, 2007).

World Organization Worked With IPR

World Intellectual Property Organization

The WIPO is a specialized agency of the United Nations. Its primary objective is to develop an international intellectual property system that is fair, accessible, and encourages innovation and creativity while safeguarding the public interest. Established in 1967 by the WIPO Convention, it operates in collaboration with member states and other international organizations to promote the protection of intellectual property worldwide. WIPO's headquarters is located in Geneva, Switzerland https://www.wipo.int/portal/en/index.html%20.

The history of WIPO dates back to 1883 when the Paris Convention was established with 14 member states. This convention created an International Bureau responsible for organizing meetings and handling administrative tasks on behalf of the member states. In 1886, the Berne Convention was introduced, focusing on copyright protection for literary and artistic works. It aimed to provide international recognition and protection of creators' rights to control and receive payment for their creative works https://www.wipo.int/treaties/en/ip/.

The International Bureau established by the Paris Convention and the one created by the Berne Convention merged in 1893 to form the United International Bureau for the Protection of Intellectual Property (BIRPI). This bureau later evolved into the WIPO, which we know today (Rao, 2007).

WIPO's mission is to foster a balanced and effective intellectual property system that promotes innovation, creativity, and economic development, while ensuring the broader public interest is taken into account. By facilitating international cooperation and harmonization of IP laws and regulations, WIPO plays a crucial role in shaping the global landscape of IPRs.

WIPO and WTO

In 1996, WIPO expanded its role and underscored the significance of IPRs in globalized trade by entering into a cooperation agreement with the WTO. This agreement aimed to facilitate cooperation between the two organizations in implementing the TRIPS Agreement. It involved activities such as exchanging information on laws and regulations, providing legal and technical assistance, and offering technical cooperation to developing countries.

In July 1998, a joint initiative was launched to assist developing countries in meeting their TRIPS obligations by the year 2000. This initiative aimed to support these countries in developing the necessary frameworks and capacities to effectively protect and enforce IPRs.

Currently, WIPO administers 24 treaties (three of which are joint treaties with other international organizations) and carries out a diverse range of programs and activities through collaboration with its member states and secretariat. The organization's work focuses on various aspects related to intellectual property, including promoting innovation, providing legal frameworks for protection, facilitating international cooperation, and addressing the challenges and opportunities brought about by advancements in technology and globalization.

- Harmonize national intellectual property legislation and procedures
- Provide services for international applications for industrial property rights
- Exchange intellectual property information
- Provide legal and technical assistance to developing and other countries
- Facilitate the resolution of private intellectual property disputes
- Marshal information technology as a tool for storing, accessing, and using valuable intellectual property information

Paris Convention for the Protection of Industrial Property

The Paris Union, established by the Paris Convention in 1883, consists of an Assembly and an Executive Committee. All member states that have adhered to the administrative and final provisions of the Stockholm Act (1967) are part of the Assembly. The members of the Executive Committee are elected from among the Union's members, with the exception of Switzerland, which is a member by default.

Over the years, the Paris Convention has undergone several revisions and amendments to adapt to changing circumstances. The revisions took place at

Brussels in 1900, Washington in 1911, The Hague in 1925, London in 1934, Lisbon in 1958, and Stockholm in 1967. Additionally, the convention was amended in 1979 to address emerging needs.

The Paris Convention covers industrial property in its broadest sense, encompassing various aspects such as patents, trademarks, industrial designs, utility models, trade names, geographical indications, and the prevention of unfair competition. It provides a framework for the protection and regulation of these IPRs at the international level, promoting cooperation and harmonization among member states https://www.wipo.int/treaties/en/ip/.

Patent Cooperation Treaty

The Patent Cooperation Treaty (PCT) established a union, which includes an assembly. Every state that is a party to the PCT becomes a member of the assembly. The assembly has several important responsibilities, including amending the regulations associated with the treaty, adopting the biennial program and budget of the union, and setting certain fees related to the use of the PCT system.

The PCT system has experienced significant growth and development over the years. For example, in 1979, the International Bureau received 2,625 international patent applications. However, by 2003, that number had increased to 110,065. Moreover, the average number of designations per application rose from 6.66 in 1979 to 132 in 2003, highlighting the expanding scope and reach of the PCT system.

The PCT was initially concluded in 1970, then amended in 1979, and subsequently modified in 1984 and 2001. It is open to states that are party to the Paris Convention for the Protection of Industrial Property (1883). To become a party to the PCT, a state must deposit its instruments of ratification or accession with the Director General of the WIPO.

The PCT allows applicants to seek patent protection for an invention simultaneously in multiple countries through the filing of an international patent application. The application can be filed by an individual who is a national or resident of a contracting state. The applicant has the option to file with the national patent office of their contracting state or with the International Bureau of WIPO in Geneva. In some cases, if the applicant is a national or resident of a contracting state that is party to the European Patent Convention, the Harare Protocol on Patents and Industrial Designs, or the Eurasian Patent Convention, they may also file the international application with the European Patent Office (EPO), the African Regional Industrial Property Organization (ARIPO), or the Eurasian Patent Office (EAPO), respectively.

Advantages of filing under the PCT

- Simultaneous protection in multiple countries
- Extended time for national phase entry
- International search report provides preliminary patentability assessment
- Cost savings by delaying expenses for individual national applications
- Streamlined and standardized application process
- Facilitates international collaboration and knowledge sharing
- Enhances patent quality through comprehensive examination
- Secures priority date for patent rights
- Flexibility in choosing national/regional routes
- Centralized administration for administrative convenience

Overall, the PCT offers time and cost efficiencies, broader protection, improved patent quality, and streamlined processes for international patent filing.

Trade-related Aspects of Intellectual Property Rights Agreement

With the establishment of the WTO, the importance and role of intellectual property protection were solidified through the TRIPS Agreement. The TRIPS Agreement was negotiated during the Uruguay Round of the GATT treaty in 1994. Its primary objectives, as stated in the Preamble to the Agreement, align with the negotiating goals established in the TRIPS area through the 1986 Punta del Este Declaration and the 1988–89 Mid-Term Review https://www.wto.org/english/tratop_e/trips_e/trips_e.htm.

The TRIPS Agreement encompasses several general goals, including

- It crystallized the importance of intellectual property protection within the WTO.
- It was negotiated during the Uruguay Round of the GATT treaty in 1994.
- The agreement aims to reduce barriers and distortions to international trade.
- It promotes effective and adequate protection of IPRs.
- TRIPS ensures that measures to enforce IPRs do not hinder legitimate trade.
- The agreement sets minimum standards for the protection and enforcement of various IPRs.
- It covers patents, trademarks, copyrights, geographical indications, industrial designs, and trade secrets.

- TRIPS establish a predictable and consistent international intellectual property regime.
- It fosters innovation, creativity, and economic growth while respecting right holders' interests.
- TRIPS strike a balance between enforcement and fair competition.

The TRIPS Agreement, which came into effect on 1 January 1995, is to date the most comprehensive multilateral agreement on intellectual property. The areas of intellectual property that it covers are:

- Copyright and related rights (i.e., the rights of performers, producers of sound recordings and broadcasting organizations)
- Trade marks including service marks
- Geographical indications including appellations of origin
- Industrial designs
- Patents including protection of new varieties of plants
- The layout designs (topographies) of ICs
- The undisclosed information including trade secrets and test data

The TRIPS Agreement covers various intellectual property-related issues, including:

- ***Patents***: Protection and enforcement of patent rights for inventions in all fields of technology
- ***Trademarks***: Protection and enforcement of distinctive signs, such as logos and brand names, to distinguish goods and services
- ***Copyrights***: Protection and enforcement of literary, artistic, and scientific works, including music, literature, software, and films
- ***Geographical indications***: Protection of indications that identify the geographical origin of goods, such as champagne or parmesan cheese
- ***Industrial designs***: Protection and enforcement of the aesthetic aspects of products, including their shape, appearance, or pattern
- ***Trade secrets***: Protection of confidential business information and preventing unfair competition through misappropriation of trade secrets
- ***Layout designs of ICs***: Protection and enforcement of designs used in the manufacturing of ICs
- ***Undisclosed information***: Protection of undisclosed information, such as confidential business and manufacturing processes
- ***Enforcement***: Measures and procedures to enforce IPRs effectively, including civil, administrative, and criminal remedies

- ***Anti-counterfeiting and piracy***: Measures to prevent and combat counterfeiting and piracy of IPRs

These issues aim to provide minimum standards for intellectual property protection and enforcement, encouraging innovation, promoting creativity, and ensuring fair trade practices among WTO member countries.

Conclusion

Intellectual property rights play a crucial role in fostering innovation, creativity, and economic development. They provide legal protection for the intellectual creations of individuals and businesses, encouraging investment in research, development, and artistic endeavors. IPRs incentivize inventors, creators, and entrepreneurs by granting them exclusive rights to their innovations, allowing them to reap the benefits of their work. These rights contribute to the growth of industries, the advancement of technology, the dissemination of knowledge, and the protection of cultural heritage. IPRs also facilitate international trade by establishing a framework for the protection and enforcement of these rights across borders. Balancing the interests of right holders with the broader public interest is essential to ensure fair competition, access to knowledge, and cultural diversity. By upholding and respecting IPRs, societies can continue to benefit from the innovations and creative contributions that shape our world.

References

Ahuja, V.K. 2017. Law Relation to Intellectual Property Rights, 3rd ed. New Delhi: LexisNexis Publication. pp. 1-852.

India Code https://www.indiacode.nic.in [Last accessed December, 2023].

Intellectual Property India. https://ipindia.gov.in/index.htm [Last accessed December, 2023].

Monagle, C., Gonzales A.T. 2001. Biodiversity & Intellectual Property Rights: Reviewing Intellectual Property Rights in Light of the Objectives of the Convention on Biological Diversity. https://www.ciel.org/Publications/tripsmay01.PDF [Last accessed December, 2023].

Paris Convention for the Protection of Industrial Property. https://www.wipo.int/treaties/en/ip/paris/%20[Last%20accessed%20December,%202023].

Rao, M.B., Guru M. 2007. Biotechnology, IPRs and Biodiversity, 1st ed. India: Pearson Publication. pp. 1-233.

Singh, B.D. 2022. Plant Breeding Principles and Methods. New Delhi: MedTech Science Press.

Singh, P. 2009. IPR and Plant Breeders Rights. New Delhi: New Vishal Publication. pp. 1-277.

World Intellectual Property Organization. https://www.wipo.int/portal/en/index.html%20[Last%20accessed%20December,%202023].

World Intellectual Property Organization. PCT—The International Patent System. https://www.wipo.int/pct/en/ [Last accessed December, 2023].

World Intellectual Property Organization. Traditional Knowledge. https://www.wipo.int/tk/en/tk/ [Last accessed December, 2023].

World Trade Organization. TRIPS Trade-Related Aspects of Intellectual Property Rights. https://www.wto.org/english/tratop_e/trips_e/trips_e.htm [Last accessed December, 2023].

Index

D

E

F

G

H

N

O

P

Q

R

S